EMBEDDED PROCESSOR-BASED SELF-TEST

FRONTIERS IN ELECTRONIC TESTING

Consulting Editor
Vishwani D. Agrawal

Books in the series:

Embedded Processor-Based Self-Test
D. Gizopoulos
ISBN: 1-4020-2785-0
Testing Static Random Access Memories
S. Hamdioui
ISBN: 1-4020-7752-1
Verification by Error Modeling
K. Radecka and Zilic
ISBN: 1-4020-7652-5
Elements of STIL: Principles and Applications of IEEE Std. 1450
G. Maston, T. Taylor, J. Villar
ISBN: 1-4020-7637-1
Fault Injection Techniques and Tools for Embedded systems Reliability Evaluation
A. Benso, P. Prinetto
ISBN: 1-4020-7589-8
High Performance Memory Memory Testing
R. Dean Adams
ISBN: 1-4020-7255-4
SOC (System-on-a-Chip) Testing for Plug and Play Test Automation
K. Chakrabarty
ISBN: 1-4020-7205-8
Test Resource Partitioning for System-on-a-Chip
K. Chakrabarty, Iyengar & Chandra
ISBN: 1-4020-7119-1
A Designers' Guide to Built-in Self-Test
C. Stroud
ISBN: 1-4020-7050-0
Boundary-Scan Interconnect Diagnosis
J. de Sousa, P.Cheung
ISBN: 0-7923-7314-6
Essentials of Electronic Testing for Digital, Memory, and Mixed Signal VLSI Circuits
M.L. Bushnell, V.D. Agrawal
ISBN: 0-7923-7991-8
Analog and Mixed-Signal Boundary-Scan: A Guide to the IEEE 1149.4 Test Standard
A. Osseiran
ISBN: 0-7923-8686-8
Design for At-Speed Test, Diagnosis and Measurement
B. Nadeau-Dosti
ISBN: 0-79-8669-8
Delay Fault Testing for VLSI Circuits
A. Krstic, K-T. Cheng
ISBN: 0-7923-8295-1
Research Perspectives and Case Studies in System Test and Diagnosis
J.W. Sheppard, W.R. Simpson
ISBN: 0-7923-8263-3
Formal Equivalence Checking and Design Debugging
S.-Y. Huang, K.-T. Cheng
ISBN: 0-7923-8184-X
Defect Oriented Testing for CMOS Analog and Digital Circuits
M. Sachdev
ISBN: 0-7923-8083-5

EMBEDDED PROCESSOR-BASED SELF-TEST

by

DIMITRIS GIZOPOULOS
University of Piraeus, Piraeus, Greece

ANTONIS PASCHALIS
University of Athens, Athens, Greece

and

YERVANT ZORIAN
Virage Logic, Fremont, California, U.S.A.

KLUWER ACADEMIC PUBLISHERS
BOSTON / DORDRECHT / LONDON

A C.I.P. Catalogue record for this book is available from the Library of Congress.

ISBN 1-4020-2785-0 (HB)
ISBN 1-4020-2801-6 (e-book)

Published by Kluwer Academic Publishers,
P.O. Box 17, 3300 AA Dordrecht, The Netherlands.

Sold and distributed in North, Central and South America
by Kluwer Academic Publishers,
101 Philip Drive, Norwell, MA 02061, U.S.A.

In all other countries, sold and distributed
by Kluwer Academic Publishers,
P.O. Box 322, 3300 AH Dordrecht, The Netherlands.

Printed on acid-free paper

CONTENTS

LIST OF FIGURES

LIST OF TABLES

to Georgia, Dora and Rita

Preface

This book discusses self-testing techniques in embedded processors. These techniques are based on the execution of test programs aiming to lower the cost of testing for processors and surrounding blocks.

Manufacturing test cost is already a dominant factor in the overall development cost of Integrated Circuits (IC). Consequently, cost effective methodologies are continuously seeked for test cost reduction. Self-test, the ability of a circuit to test itself is a widely adopted Design for Test (DfT) methodology. It does not only contribute to the test cost reduction but also improves the quality of test because it allows a test to be performed at the actual speed of the device, to detect defect mechanisms that manifest themselves as delay malfunctions. Furthermore, self-test is a re-usable test solution. It can be activated several times throughout the device's life-cycle. The self-testing infrastructure of a chip can be used to detect latent defects that do not exist at manufacturing phases, but they appear during the chip's operating life

The application of self-testing, as well as, other testing methods, face serious challenges when the circuit under test is a processor. This is due to the fact that processor architectures are particularly sensitive to performance degradation due to extensive design changes for testability improvement. DfT modifications of a circuit, including those that implement self-testing, usually lead to area, performance and power consumption overheads that may not be affordable in a processor design. Processor testing and self-testing is a particularly challenging problem due to sophisticated complex processor structure, but it is also a very important problem that needs special attention because of the central role that processors play in every electronic system.

In this book, an emerging self-test methodology that recently captured the interest of test technologists is studied. *Software-based self-testing*, also called *processor-based self-testing*, takes advantage of the programmability of processors and allows them to test themselves with the effective execution of embedded self-test programs. Moreover, software-based self-testing takes advantage of the accessibility that processors have to all other surrounding blocks of complex designs to test these blocks as well with such self-test programs. The already established System-on-Chip design paradigm that is based on pre-designed and pre-verified embedded cores employs one or more embedded processors of different architectures. Software-based self-testing is a very suitable methodology for manufacturing and in-field testing of embedded processors and surrounding blocks.

In this book, software-based self-testing is described, as a practical, low-cost, easy-to-apply self-testing solution for processors and SoC designs. It relaxes the tight relation of manufacturing testing with high-performance, expensive IC test equipment and hence results in test cost reduction. If appropriately applied, software-based self-testing can reach a very high test quality (high fault coverage) with reasonable test engineering effort, small test development cost and short test application time.

Also, this book sets a basis for comparisons among different software-based self-testing techniques. This is achieved by: describing the basic requirements of this test methodology; focusing on the basic parameters that have to be optimized; and applying it to a set of publicly available benchmark processors with different architectures and instruction sets.

Dimitris Gizopoulos *Piraeus, Greece*
Antonis Paschalis *Athens, Greece*
Yervant Zorian *Fremont, CA, USA*

Acknowledgments

The authors would like to acknowledge the support and encouragement by Dr. Vishwani D. Agrawal, the *Frontiers in Electronic Testing* book series consulting editor. Special thanks are also due to Carl Harris and Mark de Jongh of Kluwer Academic Publishers for the excellent collaboration in the production of this book.

The authors would like to acknowledge the help and support of several individuals at the University of Piraeus, the University of Athens and Virage Logic and in particular the help of Nektarios Kranitis and George Xenoulis.

Chapter 1

Introduction

1.1 Book Motivation and Objectives

Electronic products are used today in the majority of our daily activities. Thus, they enabled efficiency, productivity, enjoyment and safety.

The Integrated Circuits (ICs) realized today consist of multiple millions of logic gates and even more memory cells. They are implemented in, very deep sub-micron (VDSM) process technologies and often consist of multiple, pre-designed entities called Intellectual Property (IP) cores. This design methodology that allowed the integration of embedded IP cores is known as Embedded Core-Based System-on-Chip (SoC) design methodology. SoC design flow supported by appropriate Computer Aided Design (CAD) tools has dramatically improved design productivity and has opened up new horizons for successful implementation of sophisticated chips.

An important role in the architecture of complex SoC is played by embedded processors. Embedded processors and other cores built around them constitute the basic functional elements of today's SoCs in embedded systems. Embedded processors have optimized design (in terms of silicon area, performance, power consumption, etc), and provide the means for the integration of sophisticated, flexible, upgradeable and re-configurable functionality of a complex SoC. In many cases, more than one embedded

Embedded Processor-Based Self-Test
D.Gizopoulos, A.Paschalis, Y.Zorian

processors exist in a SoC, each of which takes over different tasks of the system and sharing the processing workload.

Issues such as the quality of the final SoC, the reliability of the manufactured ICs, and the reduced possibility of delivering malfunctioning chips to the end users, are rapidly getting more importance today with the increasing criticality of most of electronic systems applications.

In the context of these quality and reliability requirements, complex SoC designs, realized in dense manufacturing technologies face serious problems that need special consideration. Manufacturing test of complex chips based on external Automatic Test Equipment (ATE), as a method to guarantee that the delivered chips are correctly operating, is becoming less feasible and more expensive than ever. The volume of test data that must be applied to each manufactured chip is becoming very large, the test application time is increasing and the overall manufacturing test cost is becoming the dominant part of the total chip development cost.

Under these circumstances, which are expected to get worse as circuits size shrinks and density increases, the effective migration of the manufacturing test resources from outside of the chip (ATE) to on-chip, built-in resources and thus the effective replacement of external based testing with internally executed *self-testing* is, today the test technology of choice for all SoCs in practice. Self-testing allows at-speed testing, i.e. test execution at the actual operating speed of the chip. Thus all physical faults that cause either timing miss-behavior or an incorrect binary value can be detected. Also, self-testing drastically reduces test data storage requirements and test application time, both of which explode when external, ATE-based testing is used. Therefore, the extensive use of self-testing has a direct impact on the reduction of the overall chip test cost.

Testing of processors or microprocessors, even when they are not deeply embedded in a complex system, is known to be a challenging task itself. Classical testing approaches used in other digital circuits are not adequate to the carefully optimized processor designs, because they can't reach the same efficiency as in other types of digital circuits. Also, self-test approaches, successfully used to improve the testability of digital circuits, are not very suitable for processor testing because such techniques usually add overheads in the processor's performance, silicon area, pin count and power consumption. These overheads are often not acceptable for processors which have been specifically optimized to satisfy very strict area, speed and power consumption requirements.

This book primarily discusses the special problem of testing and self-testing of embedded processors in SoC architectures, as well as the problem of testing and self-testing other cores of the SoC using the embedded processor as test infrastructure.

First, the general problem of testing complex SoC architectures is discussed and the particular problem of processor testing and self-testing is analyzed. The difficulties are revealed and the requirements for successful solutions to the problem are discussed. Then, a comprehensive review of different approaches is given and the work done so far is classified into different categories depending on the targets of each methodology or application. This part of the book serves as a comprehensive guide to the readers that want to identify particular topics of processor test and apply the most suitable approach to their problem.

After this review, the processor testing and self-testing problems are discussed considering reduction of the test cost for the processor and the overall SoC. In the case of modern, cost-effective embedded processors, the extensive application of DfT techniques is limited and, in many cases, prohibited since such processors can't afford performance degradation, high hardware overhead and increased power consumption. For these reasons, the inherent processing ability of embedded processors can be taken advantage of, for successful testing of all the processor's internal modules. Several approaches have been proposed that use the processor instruction set architecture to develop efficient self-test programs which when executed perform the testing task of the processor. This technique, known as *software-based self-testing (SBST)* adds minimal or zero overheads to the normal operation of the processor and SoC in terms of extra circuit, performance penalty and power consumption. It seems that software-based self-testing, if appropriately applied, can be considered as the ultimate solution to low-cost testing of embedded processors. The requirements of software-based self-testing and different self-test styles are presented and optimization alternatives are discussed in this book. Software-based self-testing of processors is presented as an effective test approach for cost-sensitive products. A set of experimental results on several publicly available processors illustrates the practicality of software-based self-testing.

Embedded processors used in SoC designs have excellent access to other cores of the complex chip, independent of the architecture of the SoC and the interconnect style between the embedded cores. Therefore, embedded software routines, developed for software-based self-testing, can be successfully used for testing other embedded cores of the SoC. Going even further, the existence of powerful embedded processors in a design, can be used for a wide variety of tasks apart from manufacturing test alone. These tasks include: on-line testing in the field of operation, debugging, diagnosis, etc. The last part of the book briefly discusses the use of embedded processors for self-testing of SoC blocks, again from the low-cost test point of view.

The book provides a guide to processor testing and self-testing and an analysis of low-cost, software-based self-testing of processors and processor-based SoC architectures. It also sets the framework for comparisons among different approaches of software-based self-testing focusing on the main requirements of the technique. It also reveals the practicality of the method with the experimental results presented.

1.2 Book Organization

The remaining of this book is organized in the following Chapters:

- Chapter 2 discusses the trends in modern SoC design and manufacturing. The central role of embedded processors in SoC architectures is remarked. Emphasis is given to the importance that classic processors are getting today when used as embedded processors in SoC designs.
- Chapter 3 deals with the challenges of processors and processor-based SoC testing and self-testing and focuses on the particular difficulties of processor testing and the importance of the problem. Software-based self-testing is introduced and its main benefits are presented.
- Chapter 4 consists of two parts. The first part discusses several different ways of classifying processor testing approaches, since each one focuses on specific aspects of the problem. The second part presents a comprehensive, chronological list of processor testing related research work of the recent years, giving information on the focus of each work and the results obtained. Finally, each of the presented works is linked to the one or more classifications it belongs to.
- Chapter 5 discusses the concept and details of Software-Based Self-Testing. The basic objectives and requirements of the methodology from the low-cost test point of view are analyzed. Self-test code generation styles are discussed and all steps of the methodology are detailed. Optimization of self-test programs is also discussed.
- Chapter 6 is a complement to Chapter 5, as it presents a set of experimental results showing the application of software-based self-testing to several embedded processors. Target processors are of different complexities, architectures and instruction sets and the methodology is evaluated in several special cases where its pros and cons are discussed. Efficiency of software-based self-testing is, in all cases, very high.

- Chapter 7 briefly discusses the extension of software-based self-testing to SoC architectures. An embedded processor can be used for the effective testing of other cores in the SoC. The details of the approach are discussed and a list of recent, related works from the literature is given.
- Chapter 8 concludes the book, gives a quick summary of what has been discussed in it and outlines the directions in the topic that are expected to gain importance in the near future.

Chapter 2

Design of Processor-Based SoC

2.1 Integrated Circuits Technology

Integrated circuit (IC) manufacturing technologies have reached today a maturity level which is the driving force for the development of sophisticated, multi-functional and high-performance electronic systems. According to the prediction of the 2003 International Technology Roadmap for Semiconductors (ITRS) [76], by year 2018 the half pitch[1] of dynamic random access memories (DRAM), microprocessors and application-specific ICs (ASIC) will drop to 18 nm and the microprocessor physical gate length will drop to 7 nm. The implementation of correctly operating electronic circuits in such small geometries – usually referred to as Very Deep Submicron (VDSM) manufacturing technologies – was believed, just a few years ago, to be extremely difficult, if at all possible, due to major hurdles imposed by the fundamental laws of microelectronics physics when circuit elements are manufactured in such small dimensions and with such small distances separating them. Despite these skeptical views, VDSM

[1] The *half pitch* of the first level interconnect is a measure of the technology level, calculated as ½ of the *pitch* which is the sum of widths of the metal interconnect and the width of the space between two adjacent wires.

Embedded Processor-Based Self-Test
D.Gizopoulos, A.Paschalis, Y.Zorian

technologies are successfully used today to produce high performance circuits and continue providing evidence that Moore's law is still valid.

The increasing density of ICs realized in VDSM technologies allows the integration of a very large number of transistors, either in the form of logic gates or in the form of memory cells, in a single chip. In most cases, both types of circuits (logic and memory) are combined in the same chip. Furthermore, today, a single chip can contain digital circuits as well as analog or mixed-signal ones. Multi-million transistor ICs are used nowadays not only in high-end systems and performance-demanding applications, but, practically, in almost all electronic systems that people use daily at work, at home, while traveling, etc, just to mention only some examples. More system functionality can be put in electronic devices because hardware costs per transistor are now several orders of magnitude lower than in the past and the integration of more functionality to serve the final user of the system is offered at lower costs.

Successful design and implementation of circuits of such complexity and high density have become a reality not only because microelectronics manufacturing technologies matured, but also because sophisticated tools for Electronic Design Automation (EDA) or Computer-Aided Design (CAD) emerged to cope with the design complexity of such systems. An important step forward was the development and standardization of Hardware Description Languages (HDL) [51], such as VHDL [75] and Verilog [74]. HDLs, supporting simulation and synthesis EDA tools arrive together, matured together and are continuously being optimized to help designers simulate their designs at several levels of abstraction and high simulation speeds, as well as, to quickly synthesize high level (behavioral or Register Transfer Level – RTL) descriptions into working gate-level netlists that perform the desired functionality. Such a design flow based on HDLs, simulation and synthesis increases the design productivity, allows quick prototyping and early design verification at the behavioral level. This way, design errors can be identified at early stages and effectively corrected. Therefore, the possibility for a first-time-correct design is much larger.

The synergy of VDSM technologies, EDA tools and HDLs is also supported by the System-on-Chip design paradigm, discussed in section 2.2, and altogether boost design productivity.

2.2 Embedded Core-Based System-on-Chip Design

Further improvements in design productivity have been obtained recently with the wide adoption of a new systematic design methodology, well known as the System-on-Chip (SoC) design paradigm [7], [48], [60], [79], [88], [171]. SoC architectures are designed with the use of several, embedded Intellectual Property (IP) cores coming from various origins (collectively called IP core providers). IP cores are pre-designed, pre-optimized, pre-verified design modules ready to be plugged in the SoC (in a single silicon chip) and interconnected with the surrounding IP cores, to implement the system functionality. Each IP core in a SoC architecture may deliver a different form of functionality to the system: digital cores, analog cores, mixed-signal cores, memory cores, etc. IC design efficiency is expected to speed up significantly because of the emergence of the SoC design flow as well as because of the improvements in EDA technology. The 2003 ITRS [76] predicts that the SoC design cycle will drop from today's typical 12-months cycle to a 9-months cycle in year 2018.

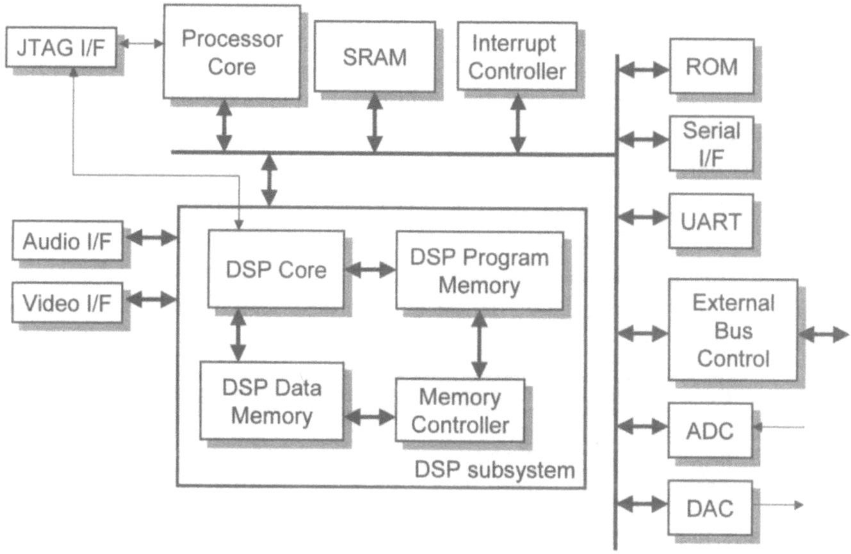

Figure 2-1: Typical System-on-Chip (SoC) architecture.

A typical SoC architecture containing representative types of cores is shown in Figure 2-1. A SoC typically consists of a number of embedded processor cores, one of which may have a central role in the architecture and others may have specialized tasks to accomplish (like the Digital Signal Processor – DSP – core of Figure 2-1, which takes over the functionality and control of the DSP subsystem of the SoC). Several memory cores (dynamic or static RAMs, ROMs, etc) are also employed in a SoC architecture, as shown in Figure 2-1, each one dedicated to a different task storing instructions, storing data or a combination of both. Other types of embedded cores are used to implement the interface between the SoC and systems out of it in a serial or parallel fashion and also other types of cores are used to interface with the analog world converting analog to digital and vice versa.

IP cores are released from IP core providers either as *soft* cores, *firm* cores or *hard* cores depending on the level of changes that the SoC designer (also called IP cores user) can make to them, and the level of transparency they come with when delivered to the final SoC integrator [60]. A *soft* core consists of a synthesizable HDL description that can be synthesized into different semiconductor processes and design libraries. A *firm* core contains more structural information, usually a gate-level netlist that is ready for placement and routing. A *hard* core includes layout and technology-dependent timing information and is ready to be dropped into a system but no changes are allowed to it [60].

Hard cores have usually smaller cost as they are final plug-and-play designs implemented in a specific technology library and no changes are allowed in them. At the opposite end, soft cores are available at HDL format and the designer can use them very flexibly, synthesizing the description using virtually any tool and any design library and thus the cost of soft cores is usually much higher than that of hard cores. A description level in between the hard and soft cores both in terms of cost and design flexibility is the firm cores case where the final SoC designer is supplied with a gate-level netlist of a design which can be altered in terms of technology library and placement/routing but not in such a flexible way as a soft core. Table 2-1 summarizes the characteristics of these three description levels of IP cores used in SoC designs.

Core category	Changes	Cost	Description
soft core	Many	High	HDL
firm core	Some	Medium	Netlist
hard core	No	Low	Layout

Table 2-1: Soft, firm and hard IP cores.

A tremendous variety of IP cores of all types and functionalities is available to SoC designers. Therefore, designers are given the great advantage to select from a rich pool of well-designed and carefully verified cores and integrate them, in a plug-and-play fashion, in the system they are designing. An idea of the variety of types of IP cores that a SoC may contain is also given in Figure 2-2.

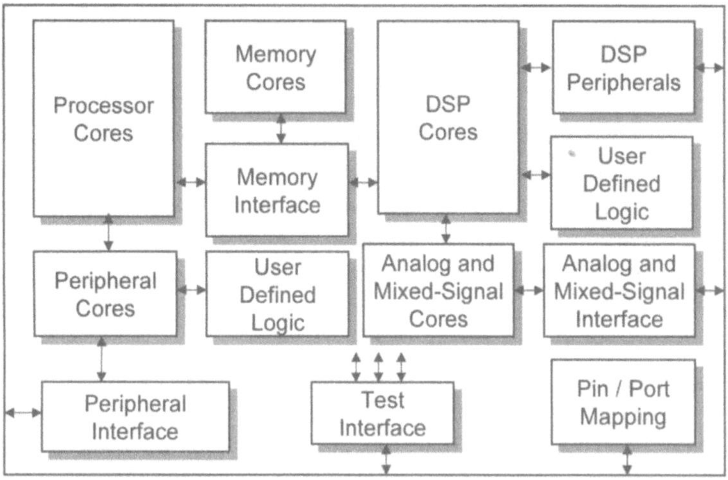

Figure 2-2: Core types of a System-on-Chip.

With the adoption of the SoC design paradigm embedded core-based ICs can be designed in a more productive way than ever and first-time-correct design is much more likely. Electronic systems designed this way have much shorter time-to-market than before and better chances for market success.

We should never forget the importance of time-to-market reduction in today's high competition electronic systems market. A product is successful if it is released to its potential users when they really need it, under the condition, of course, that it operates acceptably as expected. A "perfect" system on which hundreds of person-months have been invested may potentially fail to obtain a significant market share if it is not released on time to the target users. Therefore, successful practices at all stages of an electronic system design flow (and a SoC design flow in particular) that accomplish their mission in a quick and effective manner are always looked for. The methodology discussed in this book aims to improve one of the stages of the design flow.

We continue our discussion on the SoC design paradigm emphasizing on the key role of embedded processors in it.

2.3 Embedded Processors in SoC Architectures

Essential parts of the functionality of every SoC architecture are assigned to one or more embedded processors which are usually incorporated in a design to accomplish at least the following two tasks.

- Realization of a large portion of the system's functionality in the form of embedded code routines to be executed by the processor(s).

- Control and synchronization of the exchange of data among the different IP cores of the SoC.

The first task offers high flexibility to the SoC designers because they can use processor's inherent programmability to efficiently update, improve and revise the system's functionality just by adding or modifying existing embedded software (code and data) stored in embedded memory cores. Actually, in many situations, an updated or new product version is only a new or revised embedded software module which runs on the embedded processor of the SoC and offers new functionality to the end user of the system.

The second task that embedded processors are assigned to, offers excellent accessibility and communication from the processor to all internal cores of the SoC and, therefore, the processor can be used for several reliability-related functions of the system, the most important of them being manufacturing testing and field testing. This strong connection that an embedded processor has with all other cores of the SoC makes it an excellent existing infrastructure for the access of all SoC internal nodes, controlling their logic states and observing them at the SoC boundaries. As we will see, embedded processors can be used as an effective vehicle for low-cost self-testing of their internal components as well as other cores of the SoC.

In the majority of modern SoC architectures, more than one embedded processors exist together; the most common situation is to have two embedded processors in a SoC. For example, an embedded microcontroller (μC) or embedded RISC (Reduced Instruction Set Computer) or other processor can be used for the main processing parts of the system, while a Digital Signal Processor (DSP) can take over part of functionality of the system which is related to heavier data processing for specialized signal processing algorithms (see Figure 2-1 and Figure 2-2). In architectures where the SoC communicates with different external data channels, a separate embedded processor associated with its dedicated memory subsystem may deal with each of the communication channels while another processor can be used to coordinate the flow of data in the entire SoC.

The extensive use of embedded processors in SoC architectures of different complexities and application domains has given new life to classic processor architectures, with a word length as small as 8-bits, which were widely used in the past. Successful architectures of microcontrollers, microprocessors and DSPs were used for many years in the past, in a big variety of applications as individually packaged devices (commercial-of-the-shelf components – COTS). These classical processors are now used as embedded processor cores in complex SoC architectures and can actually boost up system's performance, while taking over simple or more complex tasks of the system's functionality.

A wide range of architectures of classical processors are now used as embedded cores in SoC designs: accumulator-based processors, stack-based processors, RISC processors, DSPs, with word sizes as small as 8 bits and as large as 64 bits. One of the most common formats that these classical processors designs appear is that of a synthesizable HDL description (VHDL or Verilog). The SoC designer (integrator) can obtain such a synthesizable processor model, synthesize it and integrate the processor in the SoC design. The SoC may be realized either using an ASIC standard cells library or an FPGA family device. Depending on the implementation technology, these processor architectures (that were considered "old" or even obsolete) can give to the SoC a processing power that is considered more than sufficient for a very large set of applications.

Furthermore, these new forms of classical processor architectures have been seriously improved in terms of performance because of the flexibility offered by the use of synthesizable HDL and today's standard cells libraries. Processors as COTS components used in boards could not be re-designed; neither could they be re-targeted in a faster technology library. On the other hand, embedded processors available in HDL format can be re-designed to meet particular needs of a SoC design as well as be re-targeted to a new technology library to obtain better performance. Even the instruction set of the processor can be altered and extended to meet the application requirements.

The reasons for the re-birth and extensive re-use of classical processor architectures in the form of embedded processor cores in SoC designs, as well as the corresponding benefits derived from this re-use are discussed in the following paragraphs.

- Classical processors have very well designed architectures, and have been extensively used, programmed, verified and tested in the past decades. The most successful of them have been able to penetrate generations of electronic products. Many algorithms have been successfully implemented as machine/assembly code for these architectures and can be effectively re-used when the new enhanced versions of the processor architectures are adopted in complex SoC designs. System development time can be saved with this re-use of existing embedded code routines. Therefore, in this case we have a dual case of re-use: *hardware re-use* (the processor core itself) and *software re-use* (the embedded routines).
- The majority of chip designers, and system architects has at least a basic knowledge of the Instruction Set Architecture (ISA) of some

classical processors and are therefore able to quickly program in their assembly language, and understand their architecture (registers, addressing modes, interrupt handling, etc). Even if a designer has experience in assembly language programming of a previous member of a processor's family, it is very easy for him/her to program in the assembly language of a later member of the processors' family. Limited man-power needs to be consumed for learning any new instruction set architecture and assembly language.

- Classical processors usually consist of a small number of gates and memory elements, and equivalently occupy small silicon area, compared with high-end RISC or CISC architectures with multiple stages of pipelining and other performance-enhancing features. Therefore, classical small processors provide an *cost-effective* solution for embedding a processing element in a SoC with small area, sufficient processing power and small electrical power consumption. This solution may of course not be suitable in all applications because of more demanding performance requirements. But for a large number of applications the performance that a classical cost-effective processor delivers is more than adequate.
- Classical processors of small word lengths, such as 8 or 16 bits, have a well-defined instruction set consisting of the most frequently used instructions. Such an instruction set can lead to small, compact programs with reduced memory requirements for code and data storage. This is a very important point in SoC architectures where the size of embedded memory components is a serious concern.

There are several examples of modern SoC architectures, used in different application domains, which include one of more classical embedded processors. This fact proves the wide adoption of classical processors as embedded cores in complex designs, in cases where the performance they offer to the system is sufficient. In all other cases, optimized, high-performance modern embedded processors are utilized to provide the system with the necessary performance for the application.

Table 2-2, Table 2-3 and Table 2-4 list a set of commercial embedded processors that are commonly used in many applications. Both categories of small, cost-effective, classical processors and also high-performance, modern processors are included in these Tables. For each embedded processor we give the company or companies that develop it and some available processor characteristics, including core type (soft, hard), core size and core performance.

Embedded Processor	Characteristics
ARClite *ARC International* `http://www.arc.com`	8-bit RISC processor. Synthesizable soft core. Less than 3500 gates for basic CPU. 40 MHz in FPGA implementation. 160 MHz in 0.18 μm ASIC process.
V8086/V186 *ARC International* `http://www.arc.com`	16-bit x86 compatible CISC processors. Synthesizable soft cores. 15,000 gates (V8086) 22,000 gates (V186). 80 MHz in 0.25 μm ASIC process.
Turbo86/Turbo186 *ARC International* `http://www.arc.com`	16-bit x86 compatible CISC processors. Synthesizable soft cores. 20,000 gates (Turbo86) 30,000 gates (Turbo186) 80+ MHz in 0.35 μm ASIC process.

C68000
CAST and Xilinx
`http://www.cast-inc.com`
`http://www.xilinx.com`
	16/32-bit microprocessor.
	Motorola MC68000 compatible.
	Hard, firm or synthesizable soft core.
	2,200 to 3,000 logic slices in various Xilinx FPGAs.
	20 MHz to 32 MHz frequency.

VZ80 *ARC International* `http://www.arc.com`	8-bit CISC processor. Z80 compatible. Synthesizable soft core. Less than 8,000 gates.
V6502 *ARC International* `http://www.arc.com`	8-bit CISC processor. 6502 compatible. Synthesizable soft core. Less than 4,000 gates.
V8-μRISC *ARC International* `http://www.arc.com`	8-bit RISC processor. Synthesizable soft core. 3,000 gates. 100 MHz in 0.25 μm ASIC process.
Y170 *Systemyde* `http://www.systemyde.com`	8-bit processor. Zilog Z80 compatible. Synthesizable soft cores. 7,000 gates.
Y180 *Systemyde* `http://www.systemyde.com`	8-bit processor. Zilog Z180 compatible. Synthesizable soft core. 8,000 gates.
DW8051 *Synopsys* `http://www.synopsys.com`	8-bit microcontroller. 803x/805x compatible. Synopsys DesignWare core Synthesizable soft core. 10,000 to 13,000 gates. 250 MHz frequency.

Table 2-2: Embedded processor cores (1 of 3).

Embedded Processor	Characteristics
DW6811 *Synopsys* `http://www.synopsys.com`	8-bit microcontroller. 6811 compatible. Synopsys DesignWare core Synthesizable soft core. 15,000 to 30,000 gates. 200 MHz frequency in 0.13 µm ASIC process.
SAM80 *Samsung Electronics* `http://www.samsung.com`	8-bit microprocessor. Zilog Z180 compatible. Hard core. 0.6 µm, 0.5 µm ASIC processes.
SM8A02/SM8A03 *Samsung Electronics* `http://www.samsung.com`	8-bit microcontroller. 80C51/80C52 subset compatible. Hard core. 0.8 µm ASIC processes.
eZ80 *Zilog* `http://www.zilog.com`	8-bit microprocessors family. Enhanced superset of the Z80 family. 50 MHz frequency.
KL5C80A12 *Kawasaki LSI* `http://www.klsi.com`	8-bit high speed microcontroller. Z80 compatible. 10 MHz frequency.

R8051
Altera and CAST Inc.
`http://www.altera.com`
`http://www.cast-inc.com`
	8-bit RISC microcontroller.
	Executes all ASM51 instructions.
	Instruction set of 80C31 embedded controller.
	Synthesizable soft core.
	2,000 to 2,500 Altera family FPGA logic cells.
	30 to 60 MHz frequency.

C8051 *Evatronix* `http://www.evatronix.pl`	8-bit microcontroller. Executes all ASM51 instructions. Instruction set of 80C31 embedded controller. Synthesizable soft core. Less than 10K gates depending on technology. 80 MHz in 0.5 µm ASIC process 160 MHz in 0.25 µm ASIC process.

DF6811CPU
Altera and Digital Core Design
`http://www.altera.com`
`http://www.dcd.com.pl`
	8-bit microcontroller CPU.
	Compatible with 68HC11 microcontroller.
	Synthesizable soft core.
	2,000 to 2,300 Altera family FPGA logic cells.
	40 to 73 MHz frequency.

MIPS32™ M4K™ *MIPS* `http://www.mips.com`	32-bit RISC microprocessor core of MIPS32™ architecture. Synthesizable soft core. 300 MHz typical frequency in 0.13 µm process. 0.3 to 1.0 mm^2 core size.

Table 2-3: Embedded processor cores (2 of 3).

Embedded Processor	Characteristics
FlexCore MIPS32 4Kec *LSI Logic* `http://www.lsilogic.com`	32-bit RISC CPU core of MIPS32™ architecture. Hard core. 167 MHz in 0.18 μm ASIC process. 200 MHz in 0.11 μm ASIC process.
ARM7TDMI *ARM* `http://www.arm.com`	32-bit RISC CPU core of ARM v4T architecture. Hard core. 133 MHz frequency in 0.13 μm ASIC process. 0.26 mm^2 core size.
ARM7TDMI-S *ARM* `http://www.arm.com`	32-bit RISC CPU core of ARM v4T architecture. Synthesizable soft core. 100 to 133 MHz frequency in 0.13 ASIC process. 0.32 mm^2 core size.
MIPS 4KE *Synopsys and MIPS* `http://www.synopsys.com`	32-bit RISC CPU core of MIPS32 ™ architecture. DesignWare Star IP Core. Synthesizable soft core. 240-300 MHz frequency. 0.4 – 1.9 mm^2 core size.

TC1MP-S
Synopsys and Infineon
`http://www.synopsys.com`
`http://www.infineon.com`
	32-bit unified microcontroller-DSP processor core.
	DesignWare Star IP Core.
	Synthesizable soft core.
	166 MHz frequency in 0.18 μm ASIC process.
	200 MHz frequency in 0.13 μm ASIC process.

PowerPC 440 *IBM* `http://www.ibm.com`	32-bit superscalar RISC processor core. Hard core. 550 MHz / 1000 MIPS in 0.15 μm ASIC process. 4.0 mm^2 core size.
AT90S2313 *Atmel* `http://www.atmel.com`	8-bit AVR-based RISC microcontroller. Includes 2 Kbyte flash memory. 10 MHz frequency.
AT90S1200 *Atmel* `http://www.atmel.com`	8-bit AVR-based RISC microcontroller. Includes 1 Kbyte flash memory. 12 MHz frequency.
C68000 *Evatronix* `http://www.evatronix.pl`	16/32-bit microprocessor. Motorola MC68000 compatible. Synthesizable soft core (VHDL and Verilog).
C32025 *Evatronix* `http://www.evatronix.pl`	16-bit fixed point Digital Signal Processor. TMS320C25 compatible. Synthesizable soft core (VHDL and Verilog).
Xtensa *Tensilica* `http://www.tensilica.com`	32-bit RISC configurable processor. Up to 300 MHz on 0.13 μm.
SHARC *Analog Devices* `http://www.analog.com`	32-bit floating point DSP core. 300 Mhz/1800 MFLOPs.

Table 2-4: Embedded processor cores (3 of 3).

The information presented in Table 2-2, Table 2-3 and Table 2-4 has been retrieved by the companies' public available documentation. It was our effort to cover a wide range of representative types of embedded processors, but of course not all available embedded processors today could be listed. The intention of the above list of embedded processors is to demonstrate that classic, cost effective processors and modern, high-performance processors are equally present today in the embedded processors market.

Apparently, when the performance that a classical, small 8-bit or 16-bit processor architecture gives to the system, is not able to satisfy the particular performance requirements of a specific application, other solutions are always available, such as the high-end RISC embedded processors or DSPs which can be incorporated in the design (several such processors are listed in the Tables of the previous pages). The high performance of modern processor architectures, enriched with deep, multi-stage pipeline structures and complex performance-enhancing circuits, is able to meet any demanding application needs (communication systems, industrial control systems, medical applications, transportation and others).

As a joint result of the recent advances in very deep submicron manufacturing technologies and design methodologies (EDA tools, HDLs and SoC design methodology), today's complex processor-based SoC devices offer complex functionality and high performance that is able to meet the needs of the demanding users of modern technology.

Unfortunately, the sophisticated functionality and the high performance of electronic systems are not offered at zero expense. Complex modern systems based on embedded processors and realized as SoC architectures have many challenges to face and major hurdles to overcome. Many of these challenges are related to the design phases of the system and others are related to the tasks of:

- verifying the circuit's correct design;
- testing the circuit's correct manufacturing; and
- testing the circuit's correct operation in the field.

These tasks have always been difficult and time consuming, even when electronic circuits size and complexity was much smaller than today. They are getting much more difficult in today's multi-core SoC designs, but they are also of increasing importance for the system's quality. An increasing percentage of the total system development cost is dedicated to these tasks of design verification and manufacturing testing. As a result, cost-reduction techniques for circuit testing during manufacturing and in the field of operation are gaining importance today and attract the attention of researchers. As we see in this book, the existence of (one or more) embedded processors in an SoC, although add to the chip's complexity, also provide a powerful embedded mechanism to assist and effectively perform testing of the chip at low cost.

In the next Chapter we discuss testing and testable design issues of embedded processors and modern processor-based SoC architectures.

Chapter 3

Testing of Processor-Based SoC

3.1 Testing and Design for Testability

The problem of testing complex SoC architectures has attracted researchers' interest the recent years because it is a problem of increasing difficulty and importance for the electronic circuits development community.

Testing, in the electronic devices world, is the systematic process used to make sure that an IC has been correctly manufactured and is free of defects. This correctness is verified with the application of appropriate inputs (called test patterns or test vectors) and the observation of the circuit's response which should be equal to the expected one, previously known from simulation. This process is called manufacturing testing and is performed before the IC is released for mounting to larger electronic systems. Manufacturing testing is applied once to each manufactured IC and occupies a significant part of the IC development cost.

The testing process may be also applied subsequently – after the IC has been released for use – to make sure that the IC continues to operate correctly when mounted in the final system. This is called periodic testing, in-field testing or on-line testing and is a necessary process because a correctly manufactured circuit that has been extensively tested during manufacturing and found to be defect-free can later malfunction because of several factors that appear in the field. Such factors are the aging of the

Embedded Processor-Based Self-Test
D.Gizopoulos, A.Paschalis, Y.Zorian

device as well as external factors such as excessive temperatures, vibrations, electromagnetic fields, induced particles, etc. Particles of relatively low size and energy can still be harmful in today's very deep submicron technologies even at the ground level because of the extremely small dimensions of the ICs being designed and manufactured today. Therefore, testing for the correct operation of a chip is not furthermore a one-time effort to be only applied during manufacturing. Testing must be repeated in regular intervals during the normal operation of the chip in its natural environment for the detection of *operational faults*[2].

The testing complexity of an electronic system (a chip, a board or a system) is conceptually decomposed in two parts, which are closely related. The first part is the complexity to generate sufficient tests, or test patterns, for the system under test – the *test generation complexity*. The second part is the complexity to actually apply tests – the *test application complexity*. Test generation is a one-time effort and constant cost for all identical manufactured devices while test application is a cost that must be accounted for each tested device. We elaborate on how these two parts of the testing complexity are related.

Test generation is usually performed as a combination of manual effort and EDA tools assistance. Of course, the increasing size of electronic systems has already led to increased needs for as much as possible automation of the test generation process. Nevertheless, even today, expert test engineers can be more effective than automatic tools in special situations. Test generation can be a very time consuming process and it may require a large number of repetitions and refinements, but in all cases, it is a one-time effort for a particular design. When a sufficient *fault coverage* level[3] has been reached by a sequence of test patterns, all identical ICs that will be manufactured subsequently will be tested using the same test sequence and no more test generation process is required, unless the design is changed. For complex designs, it may take a serious amount of person power and computing time to develop a test sequence, but after this sequence

[2] Operational faults in deep submicron technologies are classified into the following categories. *Permanent* faults are infinitely active at the same location and reflect irreversible physical changes. *Intermittent* appear repeatedly at the same location and cause errors in bursts only when they are active. These faults are induced by unstable or marginal hardware due to process variations and manufacturing residuals and are activated by environmental changes. In many cases, intermittent faults precede the occurrence of permanent faults. *Transient* faults appear irregularly at various locations and last short time. These faults are induced by neutron and alpha particles, power supply and interconnect noise, electromagnetic interference and electrostatic discharge.

[3] Fault coverage obtained by a set of test patterns is the percentage of the total faults of the chip that the test set can detect. Faults belong to a fault model which is an abstraction of the physical defect mechanisms.

is developed and is proven to guarantee high fault coverage, test generation process is considered successful.

The really hard part of the test generation process is not the actual time necessary for the development of a test set (it may vary from a few hours or days to several months). Rather, it is the ability of the test generation process itself to obtain high fault coverage for the particular design even after a long test generation time or with a large test set. In many situations, there are complex ASIC or SoC designs that even the most sophisticated sequential circuits test generation EDA tools can't handle. For such hard-to-test designs, sufficient fault coverage can't be obtained unless serious Design-for-Testability changes are applied to the circuit. Design-for-Testability (DfT) refers to design modifications that help test patterns to be easier applied to the circuit's internal nodes and node values to be easier observed at the circuit outputs.

DfT modifications are not always easily adopted by circuit designers and incorporated in the chip. The ultimate target of test generation is to obtain a high fault coverage with an as small as possible test set (to reduce test application time – discussed right after) with minimum DfT changes in the circuit. DfT changes are usually avoided by circuit designers because they degrade the performance and power behavior of the circuit during normal operation and increase the size of the circuit. Circuit testability must be considered as one of the most important design parameters and must be taken into account as early as possible in its design phases. After a designer has applied intelligent design techniques and has reached a circuit performance that satisfies the product requirements, it is very difficult to convince him/her that new design changes are necessary (with potential impact on the circuit size, performance and power consumption) to improve its testability (accessibility to internal nodes; easiness to apply test patterns and observe test responses). It is much easier for circuit designers to account for DfT structures early in the design process.

On the other hand, the second part of testing complexity, the test application complexity is related to the time interval that is necessary to apply a set of tests to the IC (by setting logic values to its primary inputs) and to observe its response (by monitoring the logic values of its primary outputs). The result of this process is the characterization of a chip as faulty or fault-free and its final rejection or delivery for use, respectively. Test application time for each individual chip depends on the number of test patterns applied to it and also on the frequency at which they are applied (how often a new test pattern is applied).

Test application time has a significant impact on the total chip development cost. This means that a smaller test application time leads to smaller device cost. On the other hand, a small test application time (testing with a smaller test set) may lead to relatively poor fault coverage. If a test set obtains small fault coverage, this means that only a small percentage of the faults that may exist in the circuit will be detected by the test set. The remaining faults of the fault model may exist in the circuit but they will not be detected by the applied test set. Therefore, the insufficiently tested device has a higher probability to malfunction when placed in the target system than a device which has been tested to higher fault coverage levels.

A discussion on the details of test generation and test application is given in the following paragraphs. Both tasks are becoming extremely difficult as the complexity of ICs and in particular processor-based SoCs increases.

Test generation for complex ICs can't be easily handled even by the most advanced commercial combinational and sequential circuit Automatic Test Pattern Generators (ATPG). ATPG tools can be used, of course, only when a gate level netlist of the circuit is available. The traditional fault models used in ATPG flows are the single stuck-at fault model, the transition fault model and the path delay fault model. The number of gates and memory elements (flip-flops, latches) to be handled by the ATPG is getting extremely high and in some cases relatively low fault coverage, of the selected fault model, can only be obtained after many hours and many backtracks of the ATPG algorithms. As circuit sizes increase this inefficiency of ATPG tools is becoming worse.

The difficulties that an ATPG faces in test generation have their sources in the reduced observability and controllability of the internal nodes in complex architectures. In earlier years, when today's embedded IP cores were used as packaged components in a System-on-Board design (packaged chips mounted to Printed Circuit Board – PCB), the chips inputs and outputs were easily accessible and testing was significantly simpler and easier because of this high accessibility. The transition to the SoC design paradigm and to miniaturized systems it develops, has given many advantages like high performance, low power consumption, small size, small weight, etc, but on the other side it imposed serious accessibility problems for the embedded cores and, therefore, serious testability problems for the SoC. Deeply embedded functional or storage cores in an SoC need special mechanisms for the delivery of the test patterns from SoC inputs to the core inputs and the propagation of their test responses from core outputs to the SoC boundaries for external observation and evaluation.

It is absolutely necessary that a complex SoC architecture includes special DfT structures to improve the accessibility to its internal nodes and thus improve its testability. The inclusion of DfT structures in a chip makes the test generation process for it much more easy and effective, and the required level of test quality can be obtained. We discuss alternative DfT approaches and their advantages and disadvantages when applied to a complex SoC.

Structured *scan-based* DfT approaches are employed to help reducing the complexity of test generation problem and accessibility of internal circuit nodes. Scan-based DfT links the memory elements (flip-flops, latches) of a digital circuit in one or more chains. Each of the memory elements can be given any logic value during the scan-in process (insertion of logic values in the scan chain) and the content of each memory element can be observed outside the chip during the scan-out process (extraction of the logic values out of the scan chain). The scan-in and scan-out processes can be performed in parallel: while a new test vector is scanned-in, the circuit response to the previous test vector is scanned-out. Scan-based DfT offers maximum accessibility to the circuit internal nodes and is also easily automated (mature commercial tools have been developed for scan-based DfT and are extensively used in industry). On the negative side, scan-based DfT suffers from the hardware overhead that it adds to the circuit and the excessive test application time that is due to long scan-in and scan-out intervals, particularly when the scan chains are very long. Scan-based SfT may have a full-scan or partial-scan architecture (all memory elements or a subset of them are connected in scan chains, respectively) [1], [23], [39], [170] and helps reducing the test generation process by an ATPG tool, by giving the ability to set values and observe internal circuit nodes. This way, the problem of sequential circuits testing is significantly simplified or even reduced to combinational circuit testing (when full scan is used).

Boundary scan architecture [124], a scan-like architecture at the chip boundaries, has been successfully used for years for board testing, for testing the chips boundaries and the interconnection between chips on a board. Boundary scan testing is continuously applied to chips today while its applications and usefulness are increasing.

At the SoC level, test communication between embedded cores, delivery of test patterns from SoC inputs to core inputs and transfer of test responses from core outputs to SoC outputs, are supported by the new IEEE 1500 Standard for Embedded Core Testing (SECT) which is being finalized [73].

IEEE 1500 SECT standardizes the test access mechanism for embedded core-based SoC, defines the test interface language (Core Test Language – CTL which is actually IEEE 1450.6, an extension to IEEE 1450 Standard Test Interface Language – STIL), as well as a hardware architecture for a core test wrapper to support the delivery of test patterns and the propagation of test responses of cores in a scan-based philosophy.

All the above scan-based test architectures (boundary scan at the chip periphery, full/partial scan at the chip or core level, and IEEE 1500 compliant scan-based testing of cores at SoC level) as well as other structured DfT techniques are very successful in reducing the test generation costs and efforts because of the existence of EDA tools for the automatic insertion of scan structures. Manual effort is very limited and high fault coverage is usually obtained. Furthermore, other structured DfT techniques like test point insertion (control and observation points) are widely used in conjunction with scan design to further increase circuit nodes accessibility and ease the test generation difficulties.

The major concerns and limitations of all scan-based testing approaches and in general of all structured or ad-hoc DfT approaches that make them not directly applicable to any design without serious considerations and planning are the following. As we see subsequently, processors are types of circuits where structured DfT techniques can't be applied in a straightforward manner.

- *Hardware overhead.*
  DfT modifications in a circuit (multiplexers for test point insertion, multiplexers for the modification of normal flip-flops to scan flip-flops, additional primary inputs and/or outputs, etc) always lead to substantial silicon area increase. In some cases this overhead is not acceptable, for example when the larger circuit size leads to a package change. Thus, DfT modifications may directly increase the chip development costs.
- *Performance degradation.*
  Scan based design and other DfT techniques make changes in the critical paths of a design. In all cases, at least some multiplexing stages are inserted in the critical paths. These additional delays may not be a problem in low-speed circuits when a moderate increase in the delay of the critical paths which is counterbalanced with better testability of the chip is not a serious concern. But, in the case of high-speed processors or high-performance processor-based SoCs such performance degradations, even at the level of 1% or 2% compared to the non-DfT design, may not be acceptable. Processor designs, carefully optimized to deliver high performance, of course, belong to this class of circuits which are particularly sensitive to performance impact due to DfT modifications.
- *Power consumption increase.*
  The increase of silicon area which is due to DfT modifications is also related to an increase in power consumption, a critical factor in many low-cost, power-sensitive designs. Scan-based DfT techniques are characterized by large power consumption because of the high circuit activity when test patterns are scanned

in the chain and test responses are scanned out of it. Circuit activity during the application of scan tests may be much more than the circuit activity during normal operation and may lead to peak power consumption not foreseen at the design stage. This can happen because scan tests apply to the circuit not functional input patterns that do not appear when the circuit operates in normal mode. Therefore, excessive power consumption during scan-based testing may seriously impact the manufacturing testing of an IC, as its package limits may be reached because of excessive heat dissipation.

- *Test data size (patterns and responses) and duration of test.* Scan-based testing among the structured DfT techniques is characterized by a large amount of test data: test patterns to be inserted in the scan chain and applied to the circuit, and test responses captured at the circuit/module outputs and then exported and evaluated externally. The total test application time (in clock cycles) in scan-based testing is many times larger compared with the actual number of test patterns, because of the large numbers of cycles required for the scan-in of a new test pattern and the scan-out of a captured test response. Test application time related to scan-in and scan-out phases is getting larger when the scan chains get longer.

The outcome of the discussion so far is that modern ATPG tools, supported by structured DfT strategies (scan chains, test points, etc) are meant to produce, in "reasonable" test generation time, a test set for the circuit under test that can be applied in "reasonable" test time (therefore, the test set size should be "sufficiently" small) to obtain a sufficiently high fault coverage of the targeted fault models. It is obvious that the meaning and value of the words "reasonable" and "sufficient" depends on the type of circuit under test and the restrictions of a particular application (total development cost, requested quality of the system, etc).

The level of hardware overheads and performance penalties that are allowed for DfT depends on the required test quality, and have a direct impact on the test application time itself: more DfT modifications of a circuit – thus more area overhead and more performance degradation – can lead to smaller test application time, in other words, to a more easily testable circuit. ATPG-based test generation time itself may not be a serious concern in many situations. Even if ATPG-based test generation lasts very long time – but eventually leads to a reasonable test set with acceptable fault coverage – it is no more than a one-time cost and effort. All the identical manufactured devices will be tested after manufacturing with the same set of test patterns.

As long as the ATPG-based test generation produces high quality tests with acceptable hardware and performance overheads, the primary concern is the test application time, i.e. the portion of the manufacturing phases that each IC spends being tested. Therefore, it is important to obtain a *small* test set at the expense of large test generation time, because a small test set will lead to a smaller test application time for each device.

Test application time per designed chip is a critical factor which has a certain impact on the production cycle duration, time-to-market and therefore, to some extend, its market success. Scan-based tests and other DfT supported test flows may lead to very large test sets that, although reach high fault coverage and test quality levels, consist of enormous amounts of test data for test pattern application and test response evaluation. This problem has already become very severe in complex SoC architectures where scan-based testing consists of: (a) core level tests that are applied to the core itself in a scan fashion (each core may include many different scan chains); and (b) SoC level tests that are used to isolate a core (again by scan techniques) and initiate the testing of the core or test the interconnection between the cores. The sizes of scan-in and scan-out data for such complex architectures have pushed the limits of modern Automatic Test Equipment (ATE, Figure 3-1) traditionally used for external manufacturing testing of ICs, because the memory capacity of ATE is usually not enough to store such huge amounts of test data.

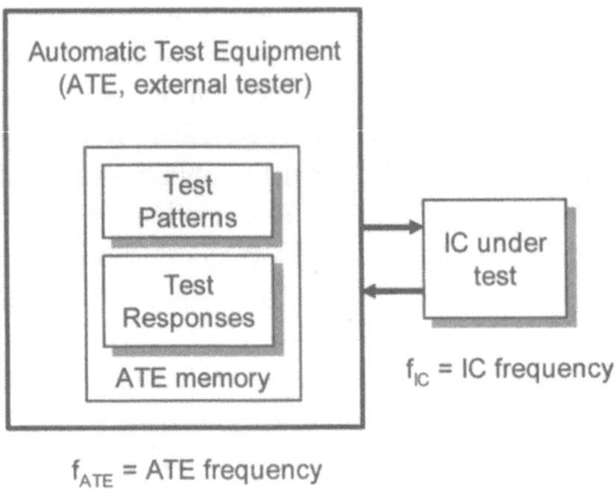

Figure 3-1: ATE-based testing.

ATE is the main mechanism with which high-volume electronic chips are tested after manufacturing. "Tester" or "external tester" are terms also used for ATE. An IC under test receives the test patterns previously stored in the tester memory and operates under this test input. The IC response to each of the test patterns is captured by the tester, stored back in the tester memory and finally compared with the known, correct response. Subsequently, the next test pattern is applied to the IC, and the process is repeated until all patterns of the test set stored in the tester memory have been applied to the chip.

The idea of ATE-based chip testing is outlined in Figure 3-1. The tester operates in a frequency denoted as f_{ATE} and the chip has an operating frequency (when used in the field mounted to the final system) denoted as f_{IC}. This means that the tester is able to apply a new test pattern at a maximum rate of f_{ATE} while the IC is able to produce correct responses when receiving new inputs at a maximum rate of f_{IC}. The relation between these two frequencies is a critical factor that determines both the *quality* of the testing process with external testers and also the test application time and subsequently the *test cost*. The relation between these two frequencies is taken in serious consideration in all cases, independently of the quality and cost of the utilized ATE (high-speed, high-cost ATE or low-speed, low-cost ATE). We will study this relation further in this book when the use of low-cost ATE in the context of software-based self-testing will be analyzed.

The essence of the relation between the tester and the chip frequencies is that if we want to execute a high quality testing to a chip and detect all (or most) physical failure mechanisms of modern manufacturing technologies, we must use a tester with a frequency f_{ATE} which is close or equal to the actual chip frequency f_{IC}. This means, in turn, that for a high frequency chip, a very expensive, high-frequency tester must be used and this fact will increase the overall test and development cost of the IC. A conflict between test quality and test cost is apparent.

Another cost-related consideration for external testing is the size of the tester's physical memory where test patterns and test responses are stored. If the tester memory is not large enough to store all the patterns of the test set and the corresponding test responses, it is necessary to perform multiple loadings of the memory, so that the entire set of test vectors is eventually applied to each manufactured chip. A larger test set requires a larger tester memory (a more expensive tester) while multiple loadings of the tester memory mean higher testing costs.

When a multi-million transistor chip is planned to be tested during manufacturing with the use of an external tester, the size of the tester memory should be large enough to avoid the application of many loadings of new chunks of test patterns in the tester memory and the multiple unloading of test responses from it. In most cases this is really infeasible: a modern complex SoC architecture can be sufficiently tested (to obtain sufficient fault coverage and fault quality) only after several loadings of new test patterns to tester memory. These multiple loadings lead to a significant amount of time of the tester being devoted to each of the devices under test. Of course the overall per-chip test application time can be significantly reduced if parallel testing of many ICs is performed but this requires the availability of even more expensive testers with this capability.

Even the use of sophisticated ATPG tools for the cores of a complex SoC and the utilization of DfT strategies (even with high hardware overhead and performance impact) are not able to significantly reduce the amount of test patterns and test responses below some level still consisting of many load-apply-store sessions in the high-end tester. Therefore, the test application time of complex SoC architectures tends to be excessive and this fact has a direct impact on the test costs and total chip development costs.

The time that an IC spends during testing under the control of an external tester adds to its total manufacturing time and final cost. Only high-end, expensive ATE of our days consisting of a huge number of channels, and a very high capacity memory for test patterns and test responses storage, and operating in very high frequencies, are capable to face the testing requirements of modern complex SoC architectures.

When a complex chip design has to be tested by an external ATE with test patterns generated by ATPG tools and possibly applied in a scan-based fashion, several factors must be taken under serious considerations. We discuss these factors in the following paragraphs in a summary of the analysis so far. The bottlenecks created by these considerations lead to new testing strategies, such as *self-testing*, that will be discussed right after. An updated discussion of all recent challenges for test technology can be found in the International Technology Roadmap for Semiconductors [76].

- *Test cost: test data volume.*
  Not all electronic applications and not all IC designs can afford the high cost of a high-end IC tester with high memory capacity and high frequency. Comprehensive testing of an IC requires the application of a sufficiently large test set in such a tester. A wide variety of IC design flows can only apply a moderate complexity test strategy to obtain a "sufficient" level of test quality at an as low as possible test development and test application cost. The test strategy followed in each design and the overall test cost that can be afforded (including the DfT cost, the test generation cost and the test application cost: use of the tester) depends on the production volume. If the volume is high enough then higher test costs are reasonable, since they are shared among the large number of chips produced and lead to small per-chip test cost; if, on the other side, volume is low then a high-end ATE-based solution is not a cost-effective solution. A tremendous number of

IC designs today have a relatively small production volume to justify the use of an expensive tester.

- *Test quality and test effectiveness: at-speed testing.*
  Even in the case that the costs related to tester purchase, maintenance, and use are not a concern in high cost, demanding applications, the test quality level seeked in such applications can't be easily reached with external testing. This is because high fault coverage tests may not be easily developed for complex designs but also because the continuously widening gap between the tester operating frequency and the IC operating frequency does not allow the detection of a large percentage of physical failures in CMOS technology that manifest themselves as delay faults (instead of logical faults). A very large set of physical mechanisms that lead to circuits not operating in the target speeds can only be detected if the chip is tested in the actual frequency in which it is expected to operate in the target system. This is called *at-speed testing*. Even for the best ATE available at any point of time, there will always be a faster IC in which performance-related circuit malfunctions will remain undetectable[4]. Therefore, the fundamental target of manufacturing testing – detection of as many as possible physical defects that may lead the IC to malfunction – can't be met under these conditions.
- *Yield loss: tester inaccuracy.*
  Testers are external devices that perform measurements on manufactured chips, and thus they suffer from severe measurement inaccuracy problems which for the high-speed designs of our days lead to serious production yield loss. A significant set of correctly operating ICs are characterized as faulty and are rejected just due to ATE inaccuracies in the performed measurements. This part of yield loss is added to the already serious yield loss encountered in VDSM technologies because of materials imperfections and equipment misses.
- *Yield loss: overtesting.*
  Another source of further yield loss is overtesting. Scan-based testing, as well as other DfT techniques, put the circuit in a mode of operation which is substantially different from its normal mode of operation. In many cases, the circuit is tested for potential faults that, even if they exist, they will never affect the

[4] This is simply because, usually, the chips of a technology generation are used in the testers of the same generation, but these testers are used to test the chips of the next generation.

normal circuit operation. The rejection of chips that have non-functional faults (like non-functionally sensitizable path delay faults) leads to further yield loss in addition to yield losses due to tester inaccuracy.

External testing of chips relying on ATE technology is a traditional approach followed by most high-volume chip manufacturers. Lower production volumes do not justify very high testing costs and the extra problems analyzed above paved the way to the *self-testing* (or built-in self-testing – BIST) technology which is now well-respected and widely applied in modern electronic devices as it overcomes several of the bottlenecks of external, ATE-based testing. Development of effective self-testing methodologies has always been a challenging task, but it is now much more challenging than in the past because of the complexity of the electronic designs to which it is expected to be applied successfully.

In the following two subsections we focus on self-testing in both its flavors: classical hardware-based self-testing and emerging software-based (processor-based) self-testing. Software-based self-testing which is the focus of this book is analyzed in detail in subsequent Chapters.

3.2 Hardware-Based Self-Testing

Hardware-based self-testing or built-in self-testing (BIST) techniques have been proposed since several decades ago [3], [4], [9], [46] to resolve the bottlenecks that external ATE-based testing can not. Self-testing is an advanced DfT technique which is based on the execution of the testing task of a chip almost completely internally, while other DfT techniques although they modify the chip's structure they perform the actual testing externally. Self-testing does not only provide the means to improve the accessibility of internal chip nodes, like any other DfT technique, but also integrate the test pattern generation and test response collection mechanisms inside the chip. Therefore, the only necessary external action is the initiation of the self-testing execution.

The problems of external, ATE-based testing, discussed in the previous subsection, have become much more difficult in the complex modern chip architectures (ASIC or SoC) and therefore the necessity and usefulness of self-testing methodologies is today much higher than in the past when designs were much smaller and more easily testable. Several testability problems of complex chips, that justify the extensive use of self-testing, have been identified in [133] (see also [23]). The factors that make testing (in particular external testing) more difficult when circuits' complexity and size increase are:

- The increasingly high logic-to-pin (gate-to-pin or transistor-to-pin) ratio, that severely affects the ability to control and observe the logic values of internal circuit nodes.
- The operating frequencies of chips which are increasing very quickly. The gigahertz (GHz) frequency domain has been reached and devices like microprocessors with multi-GHz operating frequencies are already a common practice.
- The increasingly long test pattern generation and test application times (due to increased difficulty for test generation and the excessively large test sets).
- The extremely large amount of test data to be stored in ATE memory.
- The difficulty to perform at-speed testing with external ATE. A large population of physical defects that can only be detected at the actual operating speed of the circuit escape detection.
- The unavailability of gate-level netlists and the unfamiliarity of designers with gate-level details, which both make testability structures insertion difficult. Especially, in the SoC design era with the extensive use of hard, black-box cores, gate-level details are not easy to obtain.
- Lack of skilled test engineers that have a comprehensive understanding of testing requirements and testing techniques.

Self-testing is defined as the ability of an electronic IC to test itself, i.e. to excite potential fault sites and propagate their effect to observable locations outside of the chip (Figure 3-2). The tasks of test pattern application and test response capturing/collection are both performed by internal circuit resources and not by external equipment as in ATE-based testing. Obviously, the resources used for test pattern application and test response capturing/collection should also test themselves and faults inside them must be detected. In other words, the extra circuit that is used for self-testing must also be testable. This last requirement always adds extra difficulties to self-testing methodologies.

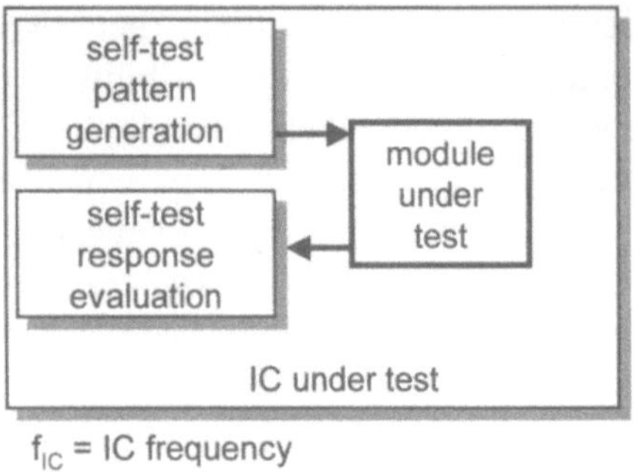

Figure 3-2: Self-testing of an IC.

In a self-testing strategy, the test patterns (as well as the expected test responses) are either stored in a special storage area on the chip (RAM, ROM) and applied during the self-testing session (we call this approach *stored patterns self-testing*), or, alternatively, they are generated by special hardware that takes over this task (we call this approach *on-chip generated patterns self-testing*). Furthermore, the actual test responses are either stored in special storage area on the chip or compressed/combined together to reduce the memory requirements. The latter case uses special circuits for test responses compaction and produces one or a few self-test *signatures*. In either case (compacted or not compacted test responses) the analysis that must eventually take place to decide if the chip is fault-free or faulty can be done inside the chip or out of it. In the extreme case that comparison with the expected correct response is done internally, a single-bit error signal comes out of the chip to denote its correct or faulty operation. The opposite extreme is the case where all test responses are extracted out of the chip for external evaluation (no compaction). The middle case (most usual in practice) is the one where a few self-test signatures (sets of compacted test responses) are collected on-chip and at the end of self-test execution are externally evaluated.

The advantages of self-testing strategies when compared to external, ATE-based testing are summarized below.

- The costs related to the purchase, use and maintenance of high-end ATE are almost eliminated when self-testing is used. There is no need to store test patterns and test responses in the tester memory, but both tasks of test application and response capturing can be performed inside the chip by on-chip resources.
- Self-testing mechanisms have a much better access to internal circuit nodes than external test mechanisms and can more likely lead to test strategies with higher fault detection capabilities. Self-testing usually obtains higher fault coverage than external

testing for a given fault model because a larger number of test patterns can be usually applied.

- Physical failure mechanisms that can only be detected when the chip operates in its actual frequency can be indeed detected with self-testing. This is the so-called *at-speed testing* requirement and obviously leads to test strategies of higher quality and effectiveness compared to others that do not detect performance-related faults (transition faults, path delay faults, etc). The superiority of self-testing at this point makes it in most cases the only choice, because in today's very deep submicron technologies an increasingly large number of failure mechanisms can only be detected by at-speed testing.
- There is no yield loss due to external testers' measurement inaccuracies simply because the chip tests itself and the test responses are captured by parts of the same silicon piece. This is a very serious concern today in the large sized chips being manufactured, which already suffer from serious yield loss problems because of their size and manufacturing process imperfections. Tester inaccuracy problems that lead to further yield loss are avoided by the utilization of appropriate self-testing techniques. The yield loss that is due to overtesting (testing for faults that can't affect the normal circuit operation) may still exist in self-testing, if the self-testing methodology tests the chip in not normal operation modes.
- The on-chip resources built for hardware-based self-testing can be re-used in later stages of the chip's life cycle, while manufacturing testing based on external equipment can not do that. A manufacturing testing strategy based on external testing is used once for each chip. On the other side, an existing self-testing mechanism that is built inside the device is an added value for the product and can be employed later during the circuit's normal operation in the field, to detect faults that can appear because of aging of the device or several environmental factors. This is known as periodic/on-line testing or in-field testing.

The use of self-testing in today's complex ICs offers significant reduction in test application time. If a large amount of test patterns is externally applied from a tester they will be applied to the chip at the tester's lower frequency and will also require multiple loadings of the tester memory which will add further delays in test application for each chip. When, on the other side, an equally large test set is applied to a chip using self-testing, not only multiple loading sessions are avoided, but also all test patterns are applied at the chip's actual frequency (usually higher than the tester frequency) and a higher test quality is achieved. In external, ATE-based testing, performance-related faults (delay/transition faults) may remain undetected because of the frequency difference between the chip and the tester, and serious portion of yield may be lost because of the tester's limited measurement accuracy. An implicit assumption made for the validity of the above statement which compares self-testing with external testing, is that the same test access mechanisms (like scan chains, test points insertion and other DfT means) are used in both cases. Only under this assumption, comparison between external testing and self-testing is fair.

Self-testing methodologies are in some situations the only feasible testing alternative strategy. These are cases where access to expensive ATE is not possible at all or when test costs associated with the use of ATE are out of question for the budget of a specific design. DfT modifications to enable self-testing may be more reasonable for the circuit designers compared with the excessive external testing costs. Many such cases exist today, for example in low-cost embedded applications where a reasonably good test methodology is needed to reach a relatively high level of test quality, but on the other side, no sophisticated solutions or expensive ATE can be used because of budget limitations. Self-testing needs only an appropriate test pattern generation flow and the design of on-chip test application and test response collection infrastructure. Hardware and performance overheads due to the employment of these mechanisms can be tailored to the specific cost limitations of a design.

Hardware-based self-testing although a proven successful testing technology for different types of small and medium sized digital circuits, is not claimed to be a panacea, as a testing strategy, for all types of architectures. A careful self-testing strategy should be planned and applied with careful guidance by the performance, cost and quality requirements of any given application. More specifically, the concerns that a test engineer should keep in mind when hardware-based self-testing is the intended testing methodology, are the following.

- *Hardware overhead.*
  This is the amount or percentage of hardware overhead devoted to hardware-based self-testing which is acceptable for the particular design. Self-testing techniques based on scan-based architectures have been widely applied to several designs. In addition to the hardware overhead that the scan design (either full or partial) adds to the circuit (regular storage elements modified into scan storage elements), or the extra multiplexing for internal

nodes access, hardware-based self-testing requires additional circuits to be synthesized and integrated in the design for the test pattern generation, the test pattern application and the test response collection tasks.

A usual tradeoff that a test engineer faces when applying a hardware-based self-testing technique is whether he/she will have the test patterns stored in a memory unit on the chip or if the test patterns will be generated by some on-chip machine specially designed for this purpose. The final decision strongly depends on the number of test patterns of the test set. If the test set consists of just a few test patterns, they can be stored in a small on-chip memory and thus the hardware overhead can be small. Otherwise, if the number of test patterns is large and their storage in an on-chip memory is not a cost-effective solution (this is the most common situation), then it is necessary that a small, "clever" on-chip sequential machine generates this large number of test patterns (in many situations necessary, for example in pseudorandom based self-testing) with a much smaller hardware overhead compared to the memory storage case. We discuss these two alternatives in the following.

- *Performance degradation.*
  This is the performance impact that the self-testing methodology is allowed to have on the circuit. In many cases, carefully designed and optimized high-performance circuits can't afford any change at all in their critical paths. In this case, hardware-based self-testing should be carefully applied to meet this strict requirement and have minimal (or, ideally, zero) impact on circuit performance during normal operation. The only way to do self-testing in such cases without any impact on the circuit's performance is to leave the critical paths of the design unaffected by the DfT changes required to apply self-testing. This is a real challenge since critical paths usually contain difficult to test parts that need DfT modifications for testability improvement.
- *Power consumption.*
  This is the additional power that the circuit may consume during hardware-based self-testing. It is a critical factor when the circuit is self-tested in a not normal operation mode, i.e. paths not activated in normal circuit operation are activated during self-testing intervals. Increased power consumption during self-testing is a usual side-effect when scan-based testing is applied and in cases that pseudorandom test patterns are used to excite the faults in a circuit. Therefore, when power consumption really

matters, scan-based self-testing and pseudorandom patterns based self-testing are not really good candidates. Power consumption is a serious concern in power-sensitive, battery-operated systems when chips are tested in the field using existing self-testing mechanisms. In this case, excessive power consumption for self-testing reduces the effective life cycle of the system's battery.

As we have already mentioned in the first paragraphs of this subsection, self-testing can be applied either using a dedicated memory that stores the test patterns as well as the test responses collected from them, or can be applied by a dedicated hardware test pattern generation machine and a separate test response analysis machine. These two different self-testing configurations are depicted in Figure 3-3 and Figure 3-4, respectively.

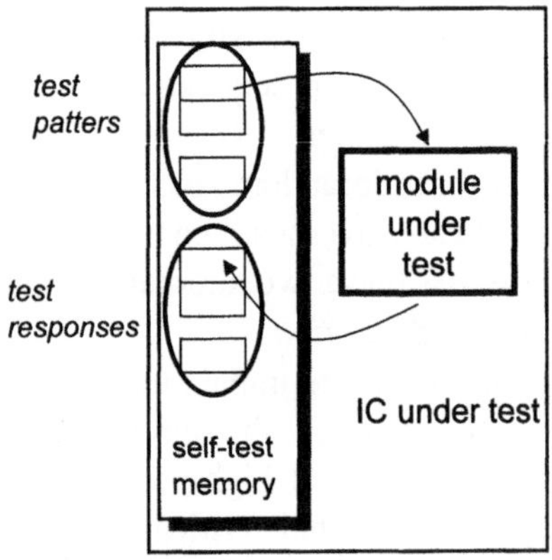

Figure 3-3: Self-testing with a dedicated memory.

In hardware-based self-testing with a dedicated memory (Figure 3-3) the actual application of the test patterns is performed by part of the circuit which reads the test patterns from the memory, applies them to the module under test and collects the test responses back to memory. The self-test memory is a dedicated memory which is only used for self-testing purposes and not the main chip's memory.

In hardware-based self-testing with dedicated hardware (Figure 3-4) the test patterns do not pre-exist anywhere, but are rather generated on-the-fly by a self-test pattern generation special circuit. Every test pattern generated by this circuit is immediately applied to the module under test and the module's response is collected and driven to another special circuit that performs test response analysis. All test reponses of the module under test are compacted by the response analyzer and a final self-test signature is eventually sent out of the chip for external evaluation.

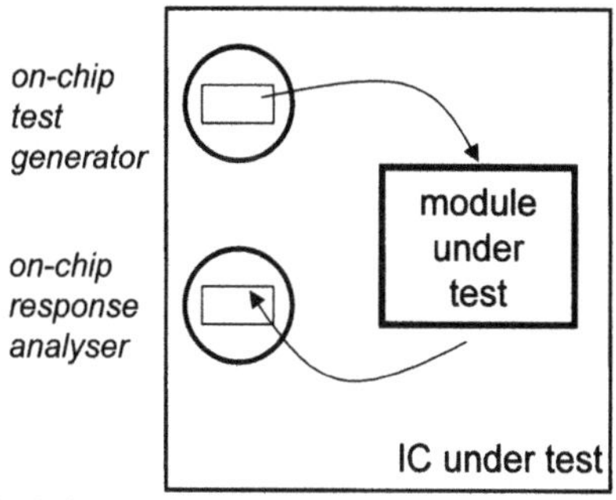

Figure 3-4: Self-testing with dedicated hardware.

A hardware-based self-testing strategy can be based on *deterministic* test patterns or *pseudorandom* test patterns. Usually, in the deterministic case the total number of test patterns is some orders of magnitude smaller than in the pseudorandom case but the on-chip test generation of the pseudorandom patterns is easier and performed by smaller circuits (called pseudorandom patterns generators) compared with the deterministic case.

Deterministic test patterns can be either previously generated by an ATPG tool (if the gate level information of the circuit under test is available) or may be, in general, previously computed or known test patterns for a module of the circuit under test. For example, several types of arithmetic circuits can be comprehensively tested with known pre-computed small test sets (even of constant size independently of the word length of the arithmetic circuit[5]). In general, deterministic test patterns are relatively few and uncorrelated/irregular. They can be stored in on-chip memory and applied during self-testing sessions as shown in Figure 3-3. The expected circuit responses are also stored in on-chip memory and the actual response of the circuit under test can be compared inside the chip with the expected fault-free responses to obtain the final pass/fail result of self-testing. The size of the dedicated on-chip memory for storing test patterns and test responses is a critical factor for deterministic self-testing and for this reason, the total number of test patterns must be rather small in order for the approach to be applicable. Such cases are rather rare in complex circuits.

On the other side, pseudorandom self-testing uses on-chip pseudorandom test patterns generators like Linear Feedback Shift Registers (LFSR) [1], [23], or Cellular Automata (CA) [1], [23]. Arithmetic circuits such as adders, subtracters and multipliers have also been shown to be successful candidates for pseudorandom pattern generation during hardware-based self-testing

[5] If a circuit that operates on n-bit operand(s) can be tested with a test set of constant size (number of test patterns) which is independent on n, we call the circuit *C-testable* and the test set a *C-test set*.

[133] because they produce pseudorandom number sequences with very good properties.

Test sets applied to circuits during pseudorandom self-testing consist of large numbers of test patterns and it usually takes quite a long test application time to reach a sufficient fault coverage (if it is at all feasible). Pseudorandom self-testing is an aggressive self-testing approach that does not rely on any algorithm for test generation but rather on the inherent ability of some circuits to be easily tested with long, random test sequences. Unfortunately, there are many circuits that are not random testable but on the contrary are strongly random-pattern resistant, i.e. high fault coverage can't be reached for them even with very long test sequences. Even in cases that relatively high fault coverage can be obtained by a first set of random patterns, the remaining faults of the circuit that should be tested to reach acceptable fault coverage are very difficult to be detected by pseudorandom test sequences.

The major advantage of pseudorandom-based self-testing is that on-chip circuits that generate the test patterns are very small and thus hardware overhead is very small too. For example, LFSRs, Cellular Automata, or Arithmetic pseudorandom patterns generators, such as accumulators, consist of simple modifications to existing circuit registers and thus have a minimum impact on circuit area.

A well-known strategy for the improvement of the efficiency of pseudorandom testing is the one where a few deterministic test patterns capable to detect the random-pattern resistant faults of a circuit are carefully embedded in long pseudorandom sequences. The resulting, "enhanced" pseudorandom test sequence reaches high fault coverage in a relatively smaller time than in pure pseudorandom testing. This improved capability is counterbalanced by the extra hardware area required for the embedding of the deterministic patterns in the pseudorandom sequence.

When the discussion comes to hardware-based self-testing of complex ASICs or embedded core-based SoC architectures, an ideal self-testing scenario is looked for. Such a scenario would be one that combines the benefits of deterministic self-testing and pseudorandom self-testing. This ideal solution would apply a relatively small set of test patterns capable to quickly obtain acceptable high fault coverage. The test patterns of the set would be also related to each other, so that they can be efficiently generated on-chip by small hardware resources and need not be stored along with their test responses in on-chip memory. In other words, the highest possible test quality and effectiveness is seeked with the smallest possible test cost.

We will have the opportunity to elaborate more on the topic of test cost through the pages of this book and show how software-based self-testing, introduced in the next subsection (also called processor-based self-testing), can be a low-cost but also high-quality self-testing methodology primarily for embedded processors but also for processor-based SoC architectures. A detailed discussion of software-based self-testing for processors is given in Chapter 5 and further.

3.3 Software-Based Self-Testing

Classical hardware-based self-testing techniques have the limitations described in the previous subsection. There are many situations where the hardware or the performance overheads that a hardware-based self-testing technique adds are not acceptable and go beyond the restrictions of the design. There are also situations where self-testing applied with the use of scan-based architectures leads to excessive power consumption because the chip is tested in a special mode of operation which is different from the normal operation for which it has been designed. It is likely that a circuit designer is reluctant to adopt even the smallest design change for hardware-based self-testing which, although it is beneficial for the circuit's testability, it will also affect the original design's performance, size and power consumption.

The existence of embedded processors in core-based SoC architectures opens the way to an alternative self-testing technique that has great potential to be very popular among circuit designers. This alternative is known as *software-based self-testing* or *processor-based self-testing* and is the focus of this book.

In software-based self-testing the test generation, test application and test response capturing are all tasks performed by embedded software routines that are executed by the embedded processor itself, instead of being assigned to specially synthesized hardware modules as in hardware-based self-testing. Processors can, therefore, be re-used as an existing testing infrastructure for manufacturing testing and periodic/on-line testing in the field. Software-based self-testing is a "natural", non-intrusive self-testing solution where the processor itself controls the flow of test data in its interior in such a way to detect its faults and no additional hardware is necessary for self-testing.

The inherent processing power that embedded processors lend to SoC designs, allows the application of any flavor and algorithm of self-testing like deterministic self-testing, pseudorandom self-testing or a combination of them. In software-based self-testing, the embedded processor executes a dedicated *software routine* or *collections of routines* that generate a sequence of test patterns according to a specific algorithm. Subsequently, the processor applies each of the test patterns of the sequence to the component under test[6], collects the component responses and finally stores them either in an unrolled fashion (each response is stored in a separate data memory word) or in a compacted form (one or more test signatures). In a multi-processor SoC design, each of the embedded processors can test itself by software routines and then they can then apply software-based self-testing to the remaining cores of the SoC.

The concept of software-based self-testing for the processor itself is illustrated in Figure 3-5. Self-test routines are stored in instruction memory and the data they need for execution are stored in data memory. Both transfers (instructions and data) are performed using external test equipment which can be as simple as a personal computer and as complex as a high-end tester. Tests are applied to components of the processor core (CPU core) during the execution of the self-test programs and test responses are stored back in the data memory.

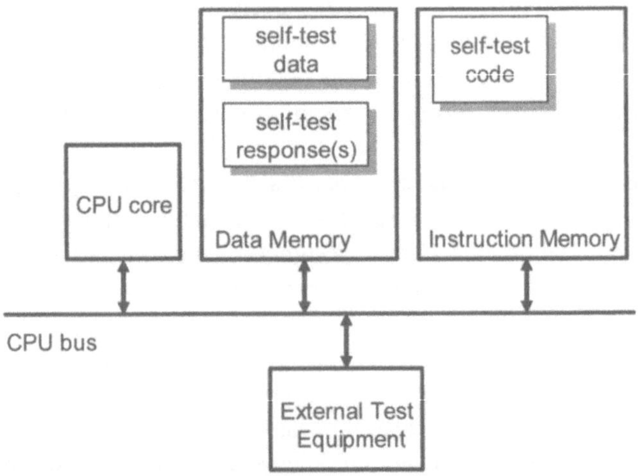

Figure 3-5: Software-based self-testing concept for processor testing.

It is implied by Figure 3-5 that the external test equipment has access to the processor bus for the transfer of self-test programs and data in the processor memory. This is not necessary in all cases. In general, there should be a mechanism that is able to download self-test program and data in the processor memory for the execution of the software-based self-testing to detect the faults of the processor.

As a step further to software-based self-testing of a processor, the concept of software-based self-testing for the entire SoC design is depicted in Figure 3-6. The embedded processor core supported with appropriately

[6] The component under test may be either an internal component of the processor, the entire processor itself or a core of the SoC other than the processor.

developed software routines, is used to test other embedded cores of the SoC. In a SoC, the embedded processor (or processors) have very good access to all cores of the SoC and therefore can access, in one way or the other, the inputs and outputs of every core. In Figure 3-6, it is again implied that there is a mechanism for the transfer of self-test code and data in the processor memory for further execution and detection of the core faults.

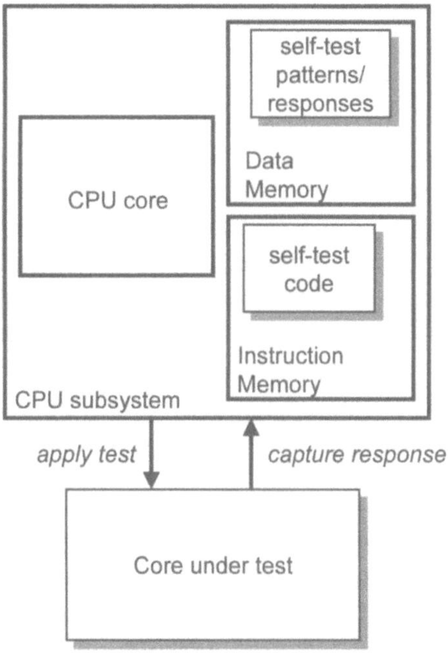

Figure 3-6: Software-based self-testing concept for testing a SoC core.

Software-based self-testing is an alternative methodology to hardware-based self-testing that has the characteristics discussed here. At this point, we outline the overall idea of software-based self-testing that clearly reveals its *low-cost* aspects. Appropriately scaled to different processor sizes and architectures, software-based self-testing is a generic SoC self-testing methodology.

- Software-based self-testing is a *non-intrusive* test methodology because it does not rely on DfT modifications of the circuit structure (the processor or the SoC). It does not add hardware and performance overheads, but on the contrary it tests the circuit using software routines executed just like all normal programs of the processor. Software-based self-testing does not affect at all the original circuit structure and does not require any modifications in the processor instruction set architecture (ISA) or the carefully optimized processor design.

- Software-based self-testing is a *low cost* test methodology because it does not rely on the use of expensive external testers and does not add any area, delay or power consumption overheads during normal operation. Low-cost, low-speed, low-pin-count testers can be perfectly utilized by software-based self-testing during manufacturing testing of a processor or an SoC, simply to download self-test routines to the processor's on-chip memory (if these routines are not already permanently stored in a flash or ROM memory), and to upload test responses or test signatures for external evaluation (if this is necessary). If the self-test program is sufficiently small then the overall test application time is minimally affected by the downloading and uploading times performed at the lower speed of the tester. The actual test application phase – i.e. the self-test code execution – will be performed at the normal operating speed of the processor which is usually much higher than the tester's speed. The usefulness of this self-testing scenario to low cost applications and low volume production flows is more than apparent. The chip manufacturing testing is not any more hooked to an expensive tester but only to low-cost equipment that simply transfers small amounts of embedded code, test patterns and test responses to and from the processor memory.
- Software-based self-testing performs *at-speed testing* of the circuit, because all test patterns are applied at the actual frequency that the circuit performs during its normal operation. This is a general characteristic of all self-testing strategies (hardware-based or software-based). Therefore, all physical failures can be detected no matter how they alter the functionality of the circuit (logic faults) or its timing behavior (delay or performance faults). Therefore, the resulting test quality is very high.
- Software-based self-testing is a *low power* test methodology because the circuit is never excited by tests that never appear during normal operation. Software-based self-testing applies test patterns that can be only delivered with normal processor instructions and not during any special test mode (like scan). Therefore, the consumed average electrical power during the execution of the self-testing programs does not differ from the average power consumption of the chip during normal operation. If the total duration of the self-testing interval is short enough

then the total power consumed during self-testing will have a minimal impact on the power behavior of the chip. This is a particularly important aspect of software-based self-testing when it is used during the entire life cycle of the circuit for on-line periodic testing in the field. Excessive power consumption in self-testing periods is a serious problem in battery-operated, hand-held and portable systems, but not if software-based self-testing is used.

- Software-based self-testing is a very *flexible* and *programmable* test strategy for complex systems because simple software code modifications or additions are sufficient to extent the testing capability to new components added to the SoC, change the processor's target fault model to another one that needs more test patterns (seeking for even higher defect coverage), or even change the purpose of execution of the routines from testing and detection of physical faults, to diagnosis and localization in order to identify the location of malfunctioning areas and cores of the design.

  The flexibility of software-based self-testing is not available in hardware-based self-testing where fixed hardware structures are only able to apply specific sets of patterns and no changes can be done after the chip is manufactured[7]. The software-based self-testing flexibility is a result of the programmability of the embedded processors and the accessibility they have to all internal components and cores of the SoC. The only factor that may block the extension and augmentation of a self-testing program is the size of the embedded memory in which it is stored. If software-based self-testing is used in manufacturing testing only, the memory size is not a problem because the normal on-chip memory (cache memory or RAM) can be used for the storage of the programs and the data. For other applications such as peridodic/on-line testing in the field, a dedicated memory (ROM, flash memory, etc) can be embedded for use of the software-based self-testing process and may need to be expanded or replaced if new or larger self-test routines are added to the system.

As we see in the next Chapter a number of hardware-based and software-based self-testing approaches have been proposed in the past along with

[7] Minor flexibility may apply to hardware-based self-testing, such as the loading of different seeds in pseudorandom patterns generators (LFSR).

several external testing approaches for processors architectures. The targets of each approach are discussed and the works are classified according to their characteristics.

We conclude the present Chapter in the following sections by giving detailed answers to three important questions that summarize the motivation of research in the area of processor self-testing and in particular software-based self-testing.

The answers that are given to these questions justify the importance of software-based self-testing for today's processor and processor-based SoCs. The questions are:

- How can software-based self-testing be an effective *test resource partitioning* technique that reduces all cost factors of manufacturing and field testing of a processor and an IC?
- Why is embedded processor testing important for the overall quality of the SoC that contains it?
- Why is embedded processor testing and processor-based testing difficult and challenging for a test engineer?

The three following subsections elaborate on these three items.

3.4 Software-Based Self-Test and Test Resource Partitioning

Test resource partitioning (TRP) is a term which was recently introduced [72] to describe the effective partitioning of several different types of test resources for IC testing (see also [25]). Test resources that must be considered for effective partitioning include:

- hardware resources (ATE hardware and built-in hardware dedicated to testing);
- time resources (test generation time and test application time);
- power resources (power consumed during manufacturing and on-line testing in the field);
- pin resources (number of pins dedicated to testing), etc.

Software-based self-testing using embedded processors is an effective TRP approach when several of these test resources are to be optimized, and this is one of its major advantages. This is the reason why we separately discuss it in this section and we elaborate on how software-based self-testing optimizes several test resources.

When *hardware* is the test resource to be optimized, it refers both to external equipment used for testing (ATE hardware and memory) and to built-in hardware dedicated to testing. Software-based self-testing relaxes the close relation of testing with high-cost ATE and reduces the external test equipment requirements to low-cost testers only for downloading of test programs to on-chip memory and uploading of final test responses from on-chip memory to the outside (the tester) for external evaluation.

In terms of special built-in hardware dedicated to test, software-based self-testing is a non-intrusive test method that does not need extra circuits just for testing purposes. On the contrary, it relies only on existing processor resources (its instructions, addressing modes, functional and control units) and *re-uses* them for testing the chip (either during manufacturing or during on-line periodic testing in the field). Therefore, compared to hardware-based self-testing that needs additional circuitry, software-based self-testing is definitely a better TRP methodology.

When *test application time* is the test resource to be optimized, software-based self-testing has a significant contribution to this optimization too. Self-testing performed by embedded software routines is a *fast* test application methodology for two simple reasons:

(a) testing is executed at the actual operating speed of the chip and not at the slower speed of the external tester (this does not only decreases the test application duration but also improves the test quality);

(b) no scan-in and scan-out cycles are required as in other hardware-based self-testing techniques that add significant delays in the testing of the device.

Test time also depends on the nature and the total number of the applied test patterns: in pseudorandom testing the number of test patterns is much larger than in the deterministic one. Software-based self-testing that applies deterministic test patterns executes in shorter time than in the case of pseudorandom-based self-testing.

When *power consumption* is the test resource under optimization, software-based self-testing never excites the processor or the SoC in other mode than its normal operating mode for which it is designed and analyzed. Other testing techniques like external testing using scan design or hardware-based self-testing with scan design again or test points insertion, test the circuit during a special mode of operation (test mode), which is completely different from normal mode. During test mode, circuit activity is much higher than in normal mode and therefore more power is consumed. It has been observed that circuit activity (and thus power consumption) during testing and self-testing can be up to three times higher than during normal mode [118], [176]. Apart from this, excessive power consumption during testing leads to energy problems in battery operated products and may also stress the limits of device packages when peak power exceeds specific thresholds.

Finally, when *pin count* is the test resource considered for optimization, software-based self-testing again provides an excellent approach. Embedded software test routines and data that just need to be downloaded from a low-cost external tester, only require a very small number of test-related chip pins. An extreme case is the use of existing JTAG boundary scan interface for this purpose. This serial downloading of embedded self-test routines may require more time than in the case of parallel download but if the routines are sufficiently small, this is not a serious problem. Self-test routines size is a metric that we will extensively consider when discussing software-based self-testing in this book.

3.5 Why is Embedded Processor Testing Important?

Embedded processors have played for a long time a key role in the development of digital circuits and are constantly the central elements in all kinds of applications. Processors are today even more important because of their increasing usage in all developed embedded systems. The answer to the question of this subsection (processor testing importance) seems to be really easy but all details must be pointed out.

The embedded processors in SoC architectures are the circuits that will execute the critical algorithms of the system and will co-ordinate the communication between all the other components/cores. The embedded processors are also expected to execute the self-test routines during manufacturing testing and in the field (on-line periodic testing), as well as, other functions like system debug and diagnosis.

As a consequence, the criticality and importance of processor testing is equivalent to the criticality and importance of its own existence in a system or a SoC. When a fault appears in an embedded processor, for example in one of its registers, then all programs that use this specific register (maybe *all* programs to be executed) will malfunction and give incorrect results. Although the fault exists only inside the processor (actually in a very small part of it), the entire system is very likely to be completely useless in the field because the system functionality expected to be executed in the processor will give erroneous output.

Other system components or SoC cores are not as critical as the processor in terms of correct functionality of the system. For example, if a memory word contains a fault, only writes and reads to this specific location will be erroneous, and this will lead to just a few programs (if any program uses this memory location at a specific point of time) to malfunction. This may not be of course a perfectly performing system but the implication of a fault in the memory module does not have so catastrophic results in the system as a fault in the embedded processor. The same reasoning is true for other cores like peripheral device controllers. If a fault exists in some peripheral device controller then the system may have trouble to access the specific device but it will be otherwise fully usable.

The task of embedded processor testing is very important because if the processor is not free of manufacturing faults it can't be used as a vehicle for the efficient software-based self-testing of the surrounding modules, and as a result the entire process of software-based self-testing will not be applicable to the particular system.

The importance of comprehensive and high quality processor testing is not at all related either with the size of an embedded processor used in a SoC architecture or its performance characteristics. A small 8-bit or 16-bit microcontroller is equally important for a successful software-based self-testing strategy when compared with a high-end 32-bit RISC processor with an advanced pipeline structure and other built-in performance-improving mechanisms. Both types of processors must be comprehensively tested and their correct operation must be guaranteed before they are used for self-testing the remaining parts of the system.

Apparently, the equal importance of all types and architectures of embedded processors for the purposes of software-based self-testing does not mean that all embedded processor architectures are tested with the same difficulty.

3.6 Why is Embedded Processor Testing Challenging?

Testing or self-testing of a processor's architecture before it can be used to perform self-testing of other components in the software-based self-testing flow, is a challenging task. From many points of view, processor architectures symbolize the universe of all design techniques [122]. All general and special design techniques are used for the components of a processor with the ultimate target to obtain the best possible performance under additional design constraints like circuit size or power consumption. For example, in many cases, the best performance is seeked under the restriction that the processor circuit size does not exceed a specific limit (probably imposed by chip packaging limitations and costs). In other cases, the limitation comes from the maximum power that can be consumed in the target applications and this is the factor that determines an upper bound to the achievable performance of a processor design. Such upper limits are usually set by the cost of the available cooling or heat removal mechanisms.

A processor is not a simple combinational unit, nor a simple finite state machine that implements a state diagram. A processor uses the best design techniques for each of its components (arithmetic units, storage elements, interconnection modules) and the best available way to make them work together under the supervision of the control unit to obtain the best performance at the instruction execution level and not the function level. This means that design optimization techniques in processors do not primarily focus on the optimization of each function (although this is useful too in most cases) but on the overall delivered performance of the processor.

Embedded processor testing and self-testing techniques are applied with several difficulties and restrictions due to optimized architecture of processors which allow only marginal (if any) changes in circuit and marginal impact on performance and power consumption. We discuss these limitations in the following. The discussion actually puts in perspective the problem of processor testing and self-testing by setting the properties/requirements that a successful processor testing technique must satisfy. The same properties/requirements are valid not only for self-testing of the processor itself but also when the processor is used to run self-testing on the remaining components of a complex SoC. Note that, software-based self-testing meets these requirements as we analyze in this book.

- Embedded processors are well-optimized designs in terms of performance and therefore the performance degradation that structured DfT (like scan design) or ad-hoc DfT techniques impose is not acceptable in most practical cases. Software-based self-testing satisfies this requirement for *zero* performance overhead, being a non-intrusive self-testing strategy that does not require any circuit modifications.
- Embedded processors are carefully optimized with respect to their size and gates count so that they occupy as small as possible silicon area when integrated in an SoC. Software-based self-testing also satisfies the requirement of virtually *zero* hardware overhead, because self-test is performed by the execution of self-test routines utilizing the processor internal resources for test generation and response capturing.
- Power consumption is a critical parameter in embedded processors designed for low-power applications used in hand-held, portable, battery operated devices. Software-based self-testing has the characteristic that it tests the processor and the overall system during normal operation mode. No special test mode is used and the average power consumption is the same as in normal circuit operation. The additional requirement is that a self-test program must have as short as possible test execution time so that the total consumed power does not have significant

affect on the battery charge available for the system. Of course this last requirement applies only for in-field test applications and on-line testing and not for manufacturing testing of the processor and the SoC.

- The self-test program used to test the embedded processor and the other SoC cores must have as small as possible size (both code and data) because it determines the amount of on-chip memory that is necessary for the storage of self-test code and data (test patterns and/or test responses). During manufacturing testing the relation between the self-test program size and the available on-chip memory determines the downloading sessions that are necessary to execute the self-testing using a low cost tester. If the self-test program size is large and multiple loadings of the memory are necessary then the test application time for manufacturing testing will be much longer. On the other side, during on-line testing the self-test program size has a direct impact on the system cost because a larger memory (ROM, flash, etc) necessary for self-test program storage will increase the system size and cost. The reduction of the self-test program size is a primary goal in software-based self-testing.
- Apart from determining the memory requirements of a software-based self-testing technique, the size of a self-test program specifies a significant portion of the total test application time (in manufacturing testing) because it is directly connected to the downloading time from the low-cost, low-speed tester to the on-chip memory. This time may be larger than the time for the execution of the self-test program because downloading is done in the low frequency of the tester while the program is executed in the higher frequency of the chip. Of course, the actual difference between the two frequencies give a better idea of which time is most important for the total test application time of the chip. As a result, the size of the self-test program is a crucial factor both for the memory requirements of the design and for the test application time and is one of the parameters that must be carefully examined and evaluated in every software-based self-testing approach.
- Another parameter related to the duration of the test application per tested chip is the execution time of the self-test program after it has been downloaded in the on-chip memory. If the difference between the tester frequency and the chip frequency is not so large then the test execution time is very important for the overall test application time. The big difference exists in cases that a

software-based self-testing approach applies a few deterministic test patterns and another approach applies a large set of pseudorandom ones. In the latter case the test execution time is much larger than in the former while, usually, smaller fault coverage is achieved.

- Last but not least of course, is the requirement for high test quality and *high fault coverage*. The target fault coverage for a software-based self-testing strategy must be as high as possible for the embedded processor itself first and then for the remaining SoC cores. An embedded processor (small, medium or large) should be tested up to a very high level of fault coverage to increase the confidence that it will operate correctly when used. This requirement is in some sense mandatory because, as we have already mentioned above, a malfunctioning processor will lead to almost all programs working incorrectly and the system being useless. Therefore, a fault that escapes manufacturing testing in a processor is much more important than a fault in a memory array or other SoC component and the obtained fault coverage for the processor itself must be larger than at the other components.

We note that processor faults that are not functionally detectable i.e. they can't be detected when circuit operates in normal mode, will not be detected by software-based self-testing. This is natural, since software-based self-testing only applies normal processor instructions for fault detection.

To give an idea of the criticality and difficulty of processor testing in software-based self-testing it is useful to outline a small but informative example. Consider a SoC consisting of an embedded processor that occupies only 20% of the total silicon area, several memory modules occupying 70% of the total area (usual situation to have large embedded memories) and other components for the remaining 10% of the SoC area.

First, it is much more important to develop a self-testing strategy that reaches a fault coverage of more than 90% or 95% for the processor than for the embedded memories although the processor size is more than three times smaller that the memories. As we explained earlier, faults in the processor that escape detection will lead to the vast majority of the programs to malfunction (independently of the exact location of the fault in the processor: it may be a register fault, a functional unit fault or a fault in the control unit). On the contrary, faults that escape detection in any of the memory cores will only lead to a small number of programs not operating correctly. Moreover, one should not forget that the overall idea of processor-based self-testing or software-based self-testing for SoC architectures can be useful and operational only when a comprehensively tested embedded processor exists in the system and it is found to be fault-free.

Secondly, apart from its importance, processor testing in our small example is much more difficult compared to memory self-testing. It is well-known that large memory arrays can be tested for very comprehensive memory fault models with small hardware machines (hardware-based self-testing, memory BIST) that produce the test patterns and compress the test responses. Memory self-testing is a successful, well-proven technology for many years now. Memory test algorithms can also be applied in the software-based self-testing framework since the embedded processor has very good access to all embedded memories and is capable to apply the necessary test sequences to them. In the above sense, therefore, although the memory arrays occupy more than three times larger area compared to the processor, they are much more easily tested than the processor.

The combination of the importance and difficulty of testing an embedded processor and subsequently use it for testing the rest of the system, reveals the need for the development of effective software-based self-testing techniques that fulfill to the maximum extent the requirements described above.

Chapter 4

Processor Testing Techniques

Intensive research has been performed since the appearance of the first microprocessor in the field of processor testing. A variety of generic methodologies as well as several ad hoc solutions has been presented in the literature. In this Chapter we provide an overview of the open literature in processor testing, putting emphasis on the different characteristics of the approaches and the requirements that each of them tries to meet.

This Chapter consists of two parts. In the first part, we discuss the characteristics of a set of different classes of processor testing techniques (external testing vs. self-testing; functional testing vs. structural testing; etc) along with the benefits and drawbacks of each one. In the second part, we briefly discuss the most important works in the area in a chronological order of publication. The Chapter concludes with the linking of the two parts, where each of the works presented in the literature is associated with the one or more classes that it belongs to aiming to provide a quick reference for those that are interested to study the area of processor testing.

4.1 Processor Testing Techniques Objectives

Each processor testing technique has different objectives and restrictions depending on the application in which it is used and the quality and budget constraints that should be met.

Embedded Processor-Based Self-Test
D.Gizopoulos, A.Paschalis, Y.Zorian

In the following sections, we elaborate on the different classes that a processor testing methodology may belong to, describing their characteristics and explaining the cases in which each of them is considered to be an effective solution. As in any other digital system testing case, there is no single solution that is applicable to all processor architectures. In many cases, more than one testing strategies are combined to provide the most efficient and suitable testing solution for a particular system's configuration. The testing strategy that is eventually applied to a processor depends on the specific processor's architecture and its instruction set, the characteristics of the particular SoC where the processor is embedded in and the system design and test cost constraints.

The classification of the variety of processor testing techniques in different classes with different objectives provides a systematic process for selecting the appropriate technique for a specific system configuration. This can be done by matching the benefits of each class with the particular requirements (test time, test cost, overheads, etc) of the particular application. In this sense, the purpose of this Chapter is to be a comprehensive survey on processor testing techniques and outline the contribution of each approach.

4.1.1 External Testing versus Self-Testing

External testing of a processor (or any IC) means that test patterns are applied to it by an external tester (ATE). The test patterns along with the expected test responses have been previously stored in the ATE memory. This is the classical manufacturing testing technique used in digital circuits. Functional test patterns previously developed for functional verification can be re-used in this scenario, potentially enhanced with ATPG patterns to increase the fault coverage.

On the other side, self-testing of a processor means that test patterns are applied to it and test responses are evaluated for correctness without the use of external ATE, but rather using internal resources of the processor. Internal resources may be either existing hardware and memory resources, or extra hardware particularly synthesized for test-related purposes (on-chip test generation and response capturing).

The benefits and drawbacks of external testing and self-testing of processors are common to those of any other digital circuit and are summarized in Table 4-1.

	Benefits	Drawbacks
External Testing	▪ Small on-chip hardware overhead ▪ Small chip performance impact	▪ Not at-speed testing ▪ High-cost ATE ▪ Only for manufacturing testing
Self Testing	▪ At-speed testing ▪ Low-cost ATE ▪ Re-usable during product's life cycle	▪ Hardware overhead ▪ Performance impact

Table 4-1: External testing vs. self-testing.

4.1.2 DfT-based Testing versus Non-Intrusive Testing

DfT techniques (ad-hoc, scan-based or other structured ones) can be used in a processor to increase its testability. Such DfT techniques can be employed either in the case that external ATE-based testing is applied to the processor or a self-testing strategy is used instead. As an example, Logic BIST (LBIST) is a classical self-testing methodology based on pseudorandom pattern generators and scan-based design. Scan based testing as well as other structured DfT techniques (like test points insertion) require many design changes and in the case of processors, although applied in some cases, it is not the selected testing style, simply because processor designers are very reluctant to adopt any major test-related design changes.

On the other hand, non-intrusive testing techniques do not require any DfT design changes in the processor and are therefore more "friendly" to the processor designers, since these techniques do not add any hardware, performance or power consumption overheads. The question with non-intrusive testing techniques is whether they are able to reach sufficient fault coverage levels since it is restricted to faults that can be detected during the normal operation of the processor (sometimes called functionally detectable faults).

The benefits and drawbacks of DfT-based testing and non-intrusive testing are summarized in the following Table 4-2.

	Benefits	Drawbacks
DfT-based testing	▪ High fault coverage ▪ Extensive use of EDA tools	▪ Non-trivial hardware, performance and power consumption overheads
Non-intrusive testing	▪ No hardware, performance or power consumption overhead	▪ Limited EDA use ▪ Low fault coverage

Table 4-2: DfT-based vs. non-intrusive testing.

4.1.3 Functional Testing versus Structural Testing

Functional testing of microprocessors and processor cores has been extensively studied the last decades. Functional testing does not try to obtain high coverage of a particular physical or structural fault model. It rather aims to test for the correctness of all known functions performed by the digital circuit. In the case of processors, functional testing aims to cover the different functions implemented by the processor's instruction set, and for this type of circuits it seems to be a very "natural" choice. Therefore, functional testing of processors needs only the instruction set architecture (ISA) information for the processor to develop test pattern sets for the processor testing and no other lower level (like gate level) model of the processor. Functional test sets may be applied either externally or internally in a self-test mode.

The drawback of functional testing is that it is not directly connected to the actual structural testability of the processor, which is related to the physical defects. The structural testability that a functional testing achieves strongly depends on the set of data (operands) which are used to test the functions of a processor. In most cases, pseudorandom operands are employed in functional testing and they lead to test sets or test programs with excessively large test application time not capable to reach high structural fault coverage. In functional testing, previously developed test sequences or test programs for design verification can be re-used for testing, and therefore test development cost is very low.

Structural testing on the other side, targets a specific structural fault model. EDA tools can be used for automatic generation of test sequences (ATPG tools) with the possible support of structured DfT techniques like scan chains or test points insertion. Structural test generation can be performed only if a gate-level model of the processor is available. If such information is available, high fault coverage can be possibly obtained for the target structural fault model with a small test set or a small test program that is executed in short time.

Table 4-3 summarizes the benefits and drawbacks of functional testing and structural testing.

	Benefits	Drawbacks
Functional testing	▪ No low-level details required ▪ Functional verification patterns can be re-used ▪ Small test development cost	▪ No relation with structural faults ▪ Low defect coverage ▪ Pseudorandom operands ▪ Long test sequences ▪ Long test programs
Structural testing	▪ EDA tools can be used ▪ High fault coverage ▪ Small test sequences ▪ Fast test programs	▪ Needs gate-level model of processor ▪ Higher test development cost

Table 4-3: Functional vs. structural testing.

Functional testing of processors has been extensively studied in the literature as it is a straightforward approach which is built upon existing test sequences of design verification and needs relatively small test development cost. The major problem with the application of functional testing is that as the complexities of processors increase, the distance between comprehensive functional testing and actual structural testability is getting longer.

4.1.4 Combinational Faults versus Sequential Faults Testing

Processor testing may focus on the detection of the faults that belong either to a combinational fault model[8] like the industry standard single stuck-at fault model, or a sequential fault model[9], i.e. a fault model which faults lead to a sequential behavior of the faulty circuit and require two-pattern tests for fault detection. Delay fault models like the path delay fault model are the most usual sequential fault models. Appropriate selection of the targeted delay faults (such as the selection of the path delay faults) must be done to reduce the ATPG and fault simulation time. ATPG and fault simulation times for sequential fault models are much larger than for combinational faults. In some cases, the number of path delay faults that are to be simulated are so many that require an excessively long fault simulation time.

EDA tools for combinational fault models (particularly for stuck-at fault model) have been used since many years ago. Their sequential fault model

8 In general, *combinational* faults alter the behavior of a circuit in such a way that combinational parts of it still behave as combinational ones but with a different – faulty – function instead of the correct one.

9 In general, *sequential* faults change the behavior of combinational parts of a circuit into a sequential one: outputs depend on the current inputs as well as on previous inputs.

counterparts are less mature but are continuously improving their performance and efficiency.

Both combinational fault testing and sequential fault testing for processors belong to the structural testing class previously defined since they both require a gate-level model of the circuit to be available for test generation and fault coverage calculation. Obviously, testing a processor for sequential (such as delay) faults using either external test patterns or built-in hardware or software routines is a testing strategy that offers higher defect coverage and higher testing quality. On the other side, combinational fault testing like stuck-at fault testing requires much smaller test sets and test programs with much shorter test execution time than testing for delay faults or other sequential faults.

Combinational testing detects the faults that change the logic behavior of the circuit and for this reason does not need to be applied at the actual frequency of the chip. Therefore, low-cost, low-speed testers can be used. On the other side, since sequential testing is executed to detect timing malfunctions, it must be executed at the actual speed of the chip.

Table 4-4 summarizes the benefits and drawbacks of combinational fault testing and sequential fault testing.

	Benefits	Drawbacks
Combinational faults testing	▪ Small test sets or test programs ▪ Short test application time ▪ Short test generation and fault simulation time ▪ EDA tools maturity	▪ Needs gate-level model of the processor ▪ Less defect coverage
Sequential faults testing	▪ Higher test quality ▪ Higher defect coverage	▪ Large test sets or test programs ▪ Long test application time ▪ Needs gate-level model of the processor ▪ Long test generation and fault simulation time ▪ Less mature EDA tools

Table 4-4: Combinational vs. sequential testing.

4.1.5 Pseudorandom versus Deterministic Testing

The fault coverage that a processor testing technique can obtain depends on the number, type and nature of the test patterns applied to the processor. Pseudorandom testing for processors can be based on *pseudorandom instruction sequences*, *pseudorandom operands* or a *combination* of these two. Pseudorandom processor testing, like in every other circuit, has the drawback that it may require excessively long test sequences to reach an acceptable fault coverage level. This is particularly true for some processor components that are random pattern resistant. Random instruction sequences are usually very unlikely to reach high fault coverage.

Despite its difficulties, pseudorandom testing of processors has been extensively studied and applied because it is a simple methodology that needs minimum engineering effort: no special test generation algorithm or tool is necessary for pseudorandom testing. Moreover, the development of pseudorandom test sequences or programs does not require a gate level model of the processor to be available. Of course, fault coverage calculations can only be done if such a model exists. Pseudorandom pattern based fault simulations are repetitively executed and may need serious amounts of time to determine an efficient *seed*[10] value for the pseudorandom sequence and a suitable *polynomial*[11] for the pseudorandom patterns generator. Appropriate selection of a seed and polynomial pair may lead to significant reductions of the test sequences that are necessary to reach the target fault coverage.

On the other hand, deterministic testing for a processor is based on either ATPG-based test sequences or other previously calculated test sets for the processor or its components. As an example, for most of the functional modules of a processor (like the ALU, multipliers, dividers, shifters, etc) there exist carefully pre-developed test sets that guarantee high fault coverage. These test sets can be applied via processor instructions to the components of the processor to obtain high fault coverage compared with the pseudorandom case. Also, high fault coverage (either for combinational or sequential fault models) can be reached with a good ATPG tool, if of course, the gate-level model of the processor is available. Therefore, the benefit of deterministic (pre-computed or ad-hoc ATPG-based) testing is the high fault coverage reached with short test sequences. On the other hand, ATPG-based testing can be done when such an EDA tool (the ATPG) is available and the success of the approach depends on the quality of the adopted tool set. Of course, when ATPG is used for combinational components of the processor the attained fault coverage is higher and is more easily obtained compared with the case of a sequential component and the use of a sequential ATPG.

Table 4-5 summarizes the benefits and drawbacks of pseudorandom testing and deterministic testing.

[10] Seed is the initial value of a pseudorandom sequence generator like an LFSR.

[11] The characteristic polynomial of an LFSR determines the sequence of pseudorandom patters that are generated and also describes the connections of its memory elements.

	Benefits	Drawbacks
Pseudorandom testing	▪ Easy development of test sequences ▪ No gate level details needed for test development	▪ Long test sequences ▪ Low fault coverage
Deterministic testing	▪ High fault coverage ▪ Short test sequences	▪ Gate level details necessary ▪ Needs special software (the ATPG) to reach a sufficient test result

Table 4-5: Pseudorandom vs. deterministic testing.

Combination of pseudorandom and deterministic testing for processors is not an unusual situation. Components of the processor that can be effectively targeted by an ATPG tool are covered by deterministic test sets while for other components a pseudorandom approach may be used. A common practice that is applied to enhance pseudorandom testing is the embedding of a few deterministic patterns in pseudorandom sequences. This method reduces the length of pseudorandom sequences and improves their detection capability.

A final remark on pseudorandom testing is that it can be re-usable and programmable in the sense that an existing pseudorandom test generator may be fed with different seeds or may be reconfigured to a different polynomial and thus it can used for the testing of different parts of the processor.

4.1.6 Testing versus Diagnosis

Processor testing techniques may be dedicated either solely to the detection of defects/faults or, additionally, to the diagnosis process and the localization of the defects/faults. Fault diagnosis is strongly connected with the manufacturing process improvement because information collected during diagnostic testing may be effectively used for the fine tuning of the manufacturing process to improve the yield of a production flow.

As in all cases, diagnostic test sets or test programs have larger complexity that their counterparts which are developed only for the detection of faults and pass/fail manufacturing testing. The actual complexity of diagnostic testing depends on the required diagnosis resolution, i.e. the cardinality of fault sets in which an individually located fault belongs to. The higher the diagnosis resolution (the smaller the fault sets) the larger the test set size that is necessary.

Table 4-6 summarizes the benefits and drawbacks of testing only methods and diagnosis methods.

	Benefits	Drawbacks
Testing	▪ Small test sequences ▪ EDA tools support	▪ Only pass/fail indication
Diagnosis	▪ High test quality ▪ Supports yield improvement	▪ Large test sequences ▪ Less EDA tools support

Table 4-6: Testing vs. diagnosis.

4.1.7 Manufacturing Testing versus On-line/Field Testing

Classical manufacturing testing techniques focus only on the detection of faults/defects that exist after chip manufacturing in the fab. A manufacturing testing strategy (ATE-based on not) is used once before the chip is released for use in the final system.

On the other side, processor testing techniques like any other IC testing technique can not only be used during manufacturing testing but also for testing the chip in the field. The latter is called on-line testing and is done while the chip is mounted in the final system and is periodically tested.

Self-testing techniques, unlike external testing ones, are excellent candidates for re-use during on-line testing because the embedded testability enhancing features and circuit structures can be re-used several times and definitely not only during the manufacturing phase of the chip. On-line periodic testing can detect faults that appear in the field because of external environmental factors or chip aging.

Table 4-7 summarizes the benefits and drawbacks of manufacturing testing and on-line testing.

	Benefits	Drawbacks
Manufacturing testing	▪ Either external or built-in	▪ May not be re-usable
On-line/field testing	▪ Re-usable during product life cycle	▪ More expensive

Table 4-7: Manufacturing vs. on-line/field testing.

4.1.8 Microprocessor versus DSP Testing

Embedded processing elements may appear in the form of either a classical von Neumann architecture with a single memory for instructions and data storage, or Harvard architecture with separate instruction and data memories. Harvard architectures are commonly used in Digital Signal Processors (DSPs) where data manipulation requirements are higher than in general purpose processors and also data transfer bandwidth must be also higher. On the other side, general purpose processors usually employ performance enhancing mechanisms like branch prediction that are not commonly used in DSPs due to the probabilistic nature of these mechanisms. Finally, both types of architectures may use relatively simple of more complex pipeline structures for performance increase.

Microprocessors and embedded processors with general purpose architecture usually have a complex control structure, compared with DSPs which usually have simpler control units but more complex and large data processing modules.

4.2 Processor Testing Literature

A relatively large number of processor testing approaches have been proposed during the last four decades. The objectives of each approach strongly depend on the type of application that the processor is used in, as well as, on the testing strategy constraints. As in any device testing problem, not all constraints and requirements can be met simultaneously.

Research activities in a particular field are sometimes clustered in small time periods where the importance of the field is high. When technological and architectural advances changed the characteristics of processors, different testing methodologies appeared. There may be also cases where research performed several years ago was found useful as an answer to modern problems.

Each of the processor testing works that we summarize below belongs to one or more of the classes of processor testing presented in the previous sections. We first list and briefly discuss each of them in chronological order and then we map each work to the classes it belongs to.

4.2.1 Chronological List of Processor Testing Research

One of the first papers in the open literature that addressed the subject of microprocessor testing was B.Willliams' paper in 1974 [169]. The internal operation of microprocessors was reviewed to illustrate problems in their testing. LSI test equipment, its requirements, its application were discussed as well as issues on wafer testing.

In 1975, M.Bilbault described and compared five methods for microprocessor testing [17]. The first method was called 'Autotest' and it assembled the microprocessor in its natural environment. A test program was running that could generate 'good' or 'bad' responses. The second method was similar to the first one, but it was based on comparison of the responses using a reference microprocessor whose output was compared with the microprocessor under test after each instruction cycle. The third method was called 'real time algorithmic' and used a prepared program to send a series of instructions to the microprocessor and to compare its response with that expected. The fourth method was called 'recorded pattern' and had two phases. In the first one, the microprocessor was simulated and the responses were recorded while in the second one the responses of the microprocessor under test were compared with the recorded responses created during the first phase. The fifth method from Fairchild was called 'LEAD' (learn, execute and diagnose) where the test program was transferred to the memory of the tester, together with the response found from a reference microprocessor, and the memory also contains all the details of the microprocessor's environment. Advantages of speed and thoroughness were claimed for 'LEAD'.

In 1975, R.Regalado introduced the concept of *user testing* of microprocessors claiming that microprocessor testing in the user's environment does not have to be difficult or laborious, even though it is inherently complex [135]. The basic concepts of the 'people' (user) oriented approach for microprocessors testing were: (1) "The test system's computer performs every function that it is capable of performing" – this is the concept of functional testing; and (2) "The communication link between the computer and people (test engineers for example) is interactive".

In 1976, E.C.Lee proposed a simple microprocessor testing technique based on microprocessor substitution [107]. Because of its low cost in hardware and software development, this technique was suitable for user testing in a simple user environment. A tester for the Intel 4004 microprocessor was described.

In 1976, D.H.Smith presented a critical study for four of the more widely accepted, during that period, methods for microprocessor testing, with a view to developing a general philosophy which could be implemented as a test with minimum effort [147]. The considered microprocessor testing methods were: actual use of microprocessor, test pattern generation based on algorithms to avoid test pattern storage, stored-response testing, and structural verification.

The pioneering work of S.Thatte and J.Abraham in functional microprocessor testing was first presented in 1978 [156]. In this paper, the task of faults detection in microprocessors, a very difficult problem because of the processors' always increasing complexity, was addressed. A general microprocessor model was presented in terms of a data processing section (simple datapath) and a control processing section, as well as, a functional fault model for microprocessors. Based on this functional fault model the authors presented a set of test generation procedures capable of detecting all considered functional faults.

S.Thatte and J.Abraham presented, in 1979 [157] and 1980 [158], respectively, test generation procedures based on a graph-theoretic model at the register transfer level (RTL). The necessary information to produce the graph model for any processor consists only of its instruction set architecture and the functions it performs. The functional testing procedures do not depend on the implementation details of the processor. The complexity of the generated tests as a function of the number of instructions of the processor is given and experimental results are reported on a Hewlett-Packard 8-bit microprocessor. A total of 8K instructions were used to obtain a 96% coverage of single stuck-at faults, which was a complete coverage of all single stuck-at faults that affect the normal operation of valid processor instructions for this benchmark. The test sequences generated by the approach presented in [158] are very long in the case of the instruction sequencing logic because of the complexity of the proposed functional test generation algorithm, while they are sufficiently short for the register decoding logic, the data path and the Arithmetic Logic Unit (ALU) of the processor.

In 1979, G.Crichton proposed a test strategy for functional testing of microprocessors where the internal logic is separated in two types: data logic and control logic in order to simplify the development of functional test vectors [34]. A practical example was presented in the form of a test program for the SAB 8080A microprocessor. The worst case of the derived functional test program lasts only 130 ms when executed at 2.5 MHz.

In 1979, C.Robach, C.Bellon, and G.Saucier proposed an application oriented test method of a microprocessor system [137]. The goal was to test the microprocessor system 'through' the application program. Thus, this program was partitioned into segments according to the hardware access points of the system and a diagnostic algorithm was proposed. The efficiency of this method with regard to the functional error hypothesis was discussed.

In 1979, P.K.Lala proposed a test method for microprocessor testing which was based on the partitioning of the microprocessor's instruction set into several instruction sets [106]. Instructions affecting the same modules inside the processor were grouped into the same instruction set. An appropriate sub-set of each instruction set was determined to form a test sequence for the microprocessor. Stuck-at faults at the address lines, data lines and the output of some internal modules were also detected by this test sequence. As an illustration, the test sequence for 8080 microprocessor was derived.

In 1980, C.Robach, G.Saucier, and R.Velazco proposed a test method for functional testing of microprocessors based upon a high level functional description of the microprocessors [139].

In 1980, C.Robach and G.Saucier considered the problem of testing a dedicated microprocessor system that performs a specific application [138]. They presented a diagnosis methodology based on a correlated analysis of the application program (modeled by a control graph) and the hardware system (modeled by data graphs).

In 1981, T.Sridhar and J.P.Hayes presented a functional testing approach for bit-sliced microprocessors [149]. A functional fault model is used in this work and complete tests are derived for bit-sliced microprocessors resembling the structure of AMD 2901 processor slice. Bit-sliced microprocessors are treated as iterative logic arrays[12] for which C-tests (tests of constant size independent on the slices/cells of the array) are sufficient to obtain complete fault coverage.

In 1981, P.Thevenod-Fosse and R.David considered the case of random testing of the data processing section of a microprocessor based on the principle that a sequence of random instructions with random data is applied simultaneously to both a processor under test and a golden processor [159]. They proposed a methodology to calculate theoretically the number of required random instructions for given instruction probabilities in user programs, based on a functional fault model for registers and operators. For the case of Motorola 6800 microprocessor they provided a program consisting of about 6,300,000 random instructions.

In 1981, B. Courtois set the basis for on-line testing of microprocessors [33]. He proposed a methodology without using massive redundancy for on-line testing that requires the periodic execution of a watch-dog and test programs, and he quantified the fault detection time (also known as on-line fault detection latency).

In 1982, M.Annaratone and M.G.Sami proposed a functional testing methodology for microprocessors [6]. They adopted microprogramming to create their functional microprocessor model starting from user available information. Their aim was to generate test procedures that detect errors associated with a functional model instead of faults, for which structural information is necessary.

In 1982, J.Jishiura, T.Maruyama, H. Maruyama, S.Kamata presented the problems raised in VLSI microprocessor testing and described a new test vector generator and timing system for external testing designed to solve these problems [82]. The test vector generator used a vertically integrated vector generation architecture to handle long arrays of test vectors, and the timing system had cross-cycle clocking, cross-cycle strobing, and multiple clocking capabilities to achieve accurate test timing.

[12] Iterative Logic Arrays (ILAs) consists of identical circuits (combinational or sequential) which are regularly interconnected in one or more dimensions.

In 1983, C.Timoc, F.Stoot, K.Wickman, L.Hess presented simulation results on processor self-testing using weighted random test patterns [161]. Input weights were optimized to obtain a shorter test sequence for the testing of the microprocessor.

In 1983, S.K.Jain and A.K.Susskind proposed that microprocessor functional testing is divided into three distinct phases: verification of the control functions and data transfer functions; verification of the data-manipulation functions; and verification of the input-output functions [77]. They considered in detail only the first of these phases. To verify control functions, they proposed a DfT technique where appropriate additional signals inside the chip were used and made observable at the terminal pins of the microprocessor chip. Complete instruction sequences executed in a test mode used to verify control functions. Also, test procedures for verifying control functions and data transfer functions were presented.

In 1983, P.Thevenod-Fosse and R.David complemented their previous work of [159], by considering the case of random testing of the control processing section of a microprocessor, as well [160].

In 1984, X.Fedi and R.David continued the work of [159] and [160] by presenting a random tester for microprocessors and comparisons between theoretical results and experimental results for random testing of Motorola 6800 microprocessors [42].

In 1984, the functional testing work of [158] was complemented by the work of D.Brahme and J.Abraham [21] which reduces the complexity of the generated tests for the processor's instruction sequencing and execution logic. A functional model based on a reduced graph is used for the microprocessor and a classification of all faults into three functional categories is given. Tests are first developed for the registers read operations and then for all remaining processor instructions. The developed tests are proposed for execution in a self-test mode by the processor itself.

In 1984, J.F.Frenzel and P.N.Marinos presented a functional testing approach for microprocessors based on functional fault models and user available information for the processor with the aim to reduce the number of tests. *N*/2-out-of-*N* codes are employed for this purpose[13] for the data words of the processor [45]. The authors applied their approach to the same hypothetical processor used in [158] and show that a smaller number of instructions are required by their approach to test the microprocessor.

In 1984, M.G.Karpovsky and R.G. van Meter presented how functional self-test techniques are used to detect single stuck-at faults in a microprocessor [86]. The derived test programs based on these techniques was of practical interest since they had a relatively small number of

[13] In general, an *m*-out-of-*n* code consists of code words which have *m* 1's and *n*-*m* 0's.

instructions with a small number of responses stored in memory. The efficiency of these techniques was demonstrated by applying them to a 4-bit microprocessor and performing single stuck-at fault simulation.

In 1984, G.Roberts and J.Masciola examined and compared the most popular techniques used in testing microprocessor-based boards and systems [140]. In particular, they concentrated on the test step that immediately follows in-circuit testing. A technique, based on memory emulation, was presented which effectively met the requirements of this test step. Application of this technique to board test, quick verification, and system test was examined, and fault detection with diagnostic capability was discussed.

In 1985, R.Koga, W.A.Kolasinski, M.T.Marra, and W.A.Hanna studied several test methods to assess the vulnerability of microprocessors to single-event upsets (SEUs) [92]. The advantages and disadvantages of each of these test methods were discussed, and the question of how the microprocessor test results can be used to estimate upset rate in space was addressed. As an application of these methods, test results and predicted upset rates in synchronous orbit were presented for a selected group of microprocessors.

In 1985, R.Velazco, H.Ziade, and E.Kolokithas presented a microprocessor test approach that allowed fault diagnosis/localization [168]. The approach was implemented on a behavioral test dedicated system: the GAPT system and it was illustrated by results obtained during testing of 80C86 microprocessors.

In 1985, R.Fujii and J.Abraham presented a methodology for functional self-test at the board level of microprocessors with integrated peripheral control modules [47]. An enhanced instruction execution fault model and new fault models for peripheral controllers were presented. Also, test program development for data compression was presented. The application of this methodology to Intel 80186 microprocessor was described.

In 1986, X.Fedi and R.David [41] completed their experimental work first presented in [42], describing a random input pattern tester for MC-6800 microprocessors. The tester was based on an efficient input pattern generator which generates the inputs with the required probability distribution. The authors illustrated the statistical properties of the test latency as a random variable. Comparisons were given between deterministic and random experiments, as well as, between experimental and theoretical results of random testing.

In 1986, P.Seetharamaiah and V.RMurthy presented a test generation global graph model at micro operation level that included architecture and organization details as parameters to be used for flexible testing of microprocessors [143]. Every microprocessor instruction was represented by its abstract execution graph, which forms a subgraph of the global graph. A tabular method was developed for a systematic ordering of the subgraphs in order of complexity, aimed at the full fault coverage of the entire processor. Based on this approach a test procedure was developed.

In 1986, B.Henshaw presented a test program for user testing of the Motorola MC68020 32-bit microprocessor [64]. The MC68020 had some features not found in earlier versions of the MC68000 family or other earlier microprocessors that affected the effective development of this test program. Test program development for user testing is based on a functional block diagram because of the limited information available for the processor architecture and structure. In this paper, the application of functional test methods to the MC68020 was examined.

In 1987, K.K.Saluja, L.Shen, and S.Y.H.Su proposed a number of algorithms for functional testing of the instruction decoding function of microprocessors [142] (earlier version of this work was presented in 1983 [141]). The algorithms were based on the knowledge of timing and control information available to users through microprocessor manuals and data sheets. They also established the order of complexity of the algorithms presented in this paper.

In 1987, P.G.Belomorski proposed a simple theoretical method for random testing of microprocessors employing the ring-wise testing concept [12]. On the basis of a functional model of a microprocessor and its faults, he dealt with the problem of establishing the length of the random test sequence needed to detect all faults with a given certainty. The method was applied to determine the random test sequence necessary for testing the MC6800 microprocessor and it was shown that the length of the random test sequence, obtained theoretically, actually covered the worst case in a pseudorandom test procedure.

In 1988, H.-P.Klug presented a microprocessor testing approach based on pseudorandom instruction sequences generated by an LFSR [91]. Pseudorandom tests generated by the LFSR are transformed into valid processor instructions. The approach is a functional testing one but with emphasis on the reduction of the test sequences size and duration. Experiments were performed on a small (5,500 transistors) execution unit of a signal processor. First, the functional testing approach of [158] was applied and it took three man months to obtain a 2,300 instructions sequence that reached 94% coverage of the single stuck-open faults for the datapath part of the unit and only 44% for the control part. On the contrary, the LFSR-based pseudorandom instruction sequences generation method that the author presented obtained only within a week, different test sequences of around 2,500 instructions that reached a 94% fault coverage for the datapath and 64% for the control part. Unfortunately, such pseudorandom-based test generation approach for larger processors will require the application of excessively large test sequences without being able to reach high fault coverage.

In 1988, L.Shen and S.Y.H.Su presented in [146] (and in a preliminary version of the work in 1984 [145]) a functional testing approach for microprocessors based on a Register Transfer Level control fault model, which they also introduced. As a first step, the read and write operations to the processor registers are tested (these operations are called the kernel) and subsequently all the processor instructions are tested using the kernel operations. The *k*-out-of-*m* codes are utilized to reduce the total number of functional tests applied to the processor. Details of the functional fault model and the procedure to derive the tests are provided in this work.

In 1989, E.-S.A.Talkhan, A.M.H.Ahmed, A.E.Salama focused on the reduction of test sequences used for microprocessor functional testing, so that as short as possible instruction sequences are developed to cover all the processor instructions [152]. Application of the method to TI's TMS32010 Digital Signal Processor shows the obtained reductions in terms of number of functional tests that must be executed.

In 1990, A.Noore and B.E.Weinrich presented a microprocessor functional testing method in which three test generation approaches were given: the modular block approach, the comprehensive instruction set approach and the microinstruction set approach [119]. These approaches were presented as viable alternatives to the exhaustive testing of all instructions, all addressing modes and all data patterns, a strategy that is getting more infeasible and impractical as processors sizes increase. This type of papers, make obvious how difficult the problem of functional testing of processors becomes as processor sizes increase. In this work, some examples of the approach application to Intel's 8085 processor are given but with no specifics on test program size and execution time or fault coverage obtained.

In 1992, A.J. van de Goor, and Th.J.W. Verhallen [58] presented a functional testing approach for microprocessors which extends the functional fault model introduced in [21] and [158] in the 80's, to cover functional units not described by the earlier functional fault model. Memory testing algorithms were integrated into the functional testing methodology to detect more complex types of faults (like coupling faults and transition faults) in the memory units of the microprocessor. The approach has been applied to Intel's i860™ processor, but no data is provided regarding the number of tests applied, and the size and execution time of the testing program.

J.Lee and J.H.Patel in 1992 [108] and 1994 [109] treated the problem of functional testing of microprocessors as a high level test generation problem for hierarchical designs. The test generation procedure was split in two phases, the path analysis phase and the value analysis phase. In path analysis, instruction sequences are developed using an instruction sequence assembling algorithm to avoid global path conflicts. The algorithm uses behavioral information of the microprocessor. In the value analysis phase, an exact value solution is computed for the module under test level and the assignment of values to all internal buses and signals for test application. Behavioral information is used to represent the internal processor architecture in a graph model. Experimental results are provided for six high level benchmarks but not for processor in total. The reported fault coverage for single stuck-at faults at the components of the high level benchmarks ranges from 48.5% up to 100.0%, depending on the benchmark and the module under test type.

In 1995, U.Bieker, P.Marwedel proposed an automatic generation approach for self-test programs of processors [18]. The approach is retargetable, i.e. it can be applied to different processor architectures since the processor specifications and instruction set are used as the algorithm input. Constraint Logic Programming (CLP) is used in this paper at the register transfer level of the processor for the generation of efficient self-test programs. The approach has been applied to four simple processor examples and no detailed structural fault coverage results are provided.

In 1996, J.Sosnowski, A.Kusmierczyk presented an interesting comparison between deterministic and pseudorandom testing of microprocessors using the Intel 80x86 processors architecture as a demonstration vehicle [148]. The drawbacks of each of the two approaches were discussed and a combined approach was claimed to be better than any of the two alone. As in other types of circuits testing, deterministic testing is limited by the unavailability of structural information of the circuit, while pseudorandom testing may require very large test sequences without obtaining sufficient fault coverage and thus must be combined with deterministic tests.

In 1996, S.M.I.Adham and S.Gupta proposed a BIST technique, termed DP-BIST, suitable for high performance DSP datapaths [2]. The BIST session for the DSP was controlled via hardware without the need for a separate test pattern generation register or test program storage. Furthermore, the BIST scenario was appropriately set-up so as to also test the register file as well as the shift and truncation logic in the datapath. The use of DP-BIST enabled at-speed testing with no performance degradation and little area overhead for the hardware test control. Besides, they showed how DB-BIST can be used as a centralized test resource to test other macros on the chip, as well as, the integration of DP-BIST with internal scan and boundary scan.

In 1997, K.Radecka, J.Rajski and J.Tyszer introduced Arithmetic Built-In Self-Test (ABIST) as an effective pseudorandom-based technique for testing the datapath of DSP cores using the functionality of existing arithmetic modules [131]. Arithmetic modules are used to generate pseudorandom tests and to compact test responses while they test themselves, other components of the DSP core and external circuits. Test application and response collection can be performed using software routines of the core.

The work of 1997 [165] by R.S.Tupuri, J.A.Abraham is a first presentation of [166] were the authors described the approach (constraint extraction, constraint-based ATPG). This first approach of [165] was relying on commercial ATPG tools that were inadequate to reach high fault coverage levels.

In 1997, K.Hatayama, K.Hikone, T.Miyazaki, H.Yamada, proposed instruction-based test generation for functional modules in processors as a constrained test generation problem [61]. An instruction-based test generation system ALPS (ALU-oriented test pattern generation system) is outlined. The approach targets functional units (ALUs) of processors, by translating module-level tests into instruction sequences taking into consideration the instruction set architecture imposed constraints. Experimental results are given in [61] for two functional units, a floating point adder and a floating point multiplier of a RISC processor (without further details on the identity of the processor and its architecture characteristics). The constrained test generation process for the floating point adder reaches 89.10% fault coverage, while for the floating point multiplier it reaches 89.15% fault coverage.

In 1998, J.Shen and J.Abraham [144] presented a functional test generation approach for processors and applied it to manufacturing testing as well as to design validation. No a priori structural fault model is considered in the approach and the information that is necessary for the test generation method (and prototype tool) to operate is the processor's instruction set and the operations that the processor performs as a response to each instruction. For the functional testing of each instruction, the approach generates a sequence of instructions that enumerates all the combinations of the instruction operation and systematically selected operands. Also, random instruction sequences are generated for groups of instructions, to exercise the instructions and propagate their effects. Experiments have been performed on two processor benchmarks: the Viper 32-bit processor [36] and GL85 which is a model of Intel's 8085 processor. For Viper, the proposed test generation method obtains 94.04% single stuck-at fault coverage. For GL85, a fault coverage of 90.20% for single stuck-at faults is obtained again as a combination of the two facets of the methodology: exhaustive testing of each instruction with all combinations of operations and operands, and random generation of instruction sequences for groups of instructions. This approach, being a functional one that relies on exhaustive and/or pseudorandom testing, applies very long instruction sequences. For example, it is reported in [144] that a total of 360,000 instruction cycles were necessary for the GL85 processor to execute a self-test program that reached a 86.7% fault coverage for single stuck-at faults.

In 1998, DSP testing using self-test programs was considered by W.Zhao and C.Papachristou [175]. This approach is based on the application of random patterns to the components of the DSP using self-test programs developed based on behavioral and structural level testability analysis. Such analysis is not performed in functional self-testing where neither information on the RTL description of the processor is required nor a specific structural fault model is considered. Experimental results are provided in [175], showing that a 94.15% single stuck-at fault coverage is obtained by this pseudorandom approach in a simple DSP designed by the authors for the experiments. No details were given regarding the size of the self-test program, the number of test patterns applied to the DSP datapath modules or the total number of clock cycles for the execution of the self-test program.

In 1999, R.S.Tupuri, A.Krishnamachary, and J.A.Abraham proposed an automatic functional constraint extraction algorithm that transforms modules of a processor by attaching virtual logic which represents the functional constraints [166]. Then, automatic test generation can be executed on the transformed components to derive tests that can be realized with processor instructions. The implemented algorithm has been applied to different modules of three processor models (Viper, DLX and ARM) and fault coverage for single stuck-at faults between 81.14% and 97.17% was reported. These fault coverages are much higher when compared with the case that ATPG is done at the processor level for the same processor components without the use of the constraint extraction of the proposed algorithm.

In 1999, K.Batcher and C.Papachristou proposed Instruction Randomization Self Test (IRST) for processor cores, a pseudorandom self-testing technique [10]. Self-test is performed with processor instructions that are randomized by a special circuitry designed out of the processor core for this purpose. IRST does not add any performance overhead to the processor and the extra hardware is relatively small compared to the processor size (3.1% hardware overhead is reported for a DLX-like RISC processor core). The obtained fault coverage for the processor core after the execution of a random instructions sequence running for 50,000 instruction cycles is 92.5%, and after the execution of 220,000 instruction cycles is 94.8% (processor size is 27,860 and it contains 43,927 single stuck-at faults).

In 2000 W.-C.Lai, A.Krstic and K.-T.Cheng proposed a software-based self-testing technique with respect to path delay faults [105]. The proposed approach is built upon constraint extraction, classification of the paths of the processor, constraint structural ATPG (thus, deterministic path delay test patterns are used) and automatic test program synthesis. The target is the detection of delay faults in functionally testable paths of the processor. The entire flow requires knowledge of the processor's instruction set, its micro-architecture, RTL netlist as well as the gate-level netlist for the identification of the functionally testable paths. Experiments have been performed for the Parwan educational processor [116] as well as for the DLX RISC processor [59]. In Parwan, as self-test program of 5,861 instructions (bytes) obtained a 99.8% coverage of all the functionally testable path delay faults, while in DLX, a self-test program of 34,694 instructions (32-bits words) obtained a 96.3% coverage of all functionally testable path delay faults.

The contribution of the work presented by L.Chen and S.Dey in 2001 [28] (preliminary version was presented in [27]) is twofold. First, it demonstrates the difficulties and inefficiencies of Logic BIST (LBIST) application to embedded processors. This is shown by applying Logic BIST to a very simple 8-bit accumulator-based educational processor (Parwan [116]) and a stack-based 32-bit soft processor core that implements the Java Virtual Machine (picoJava [127]). In both cases, Logic BIST adds more hardware overhead compared to full scan, but is not able to obtain satisfactory structural fault coverage even when a very high number of test patterns are applied. Secondly, a structural software-based self-testing approach is proposed in [28] based on the use of self-test signatures. Self-test signatures provide a compacted way to download previously prepared test patterns for the processor components into on-chip memory. The self-test signatures are expanded by embedded software routines into test patterns which are in turn applied to the processor components and test responses (either individually or in a compacted signature) are collected for external evaluation. The component test sets are either previously generated by an ATPG and then embedded in pseudorandom sequences or are generated by software implemented pseudorandom generators (LFSRs). Experimental results on the Parwan educational processor show that 91.42% fault coverage is obtained for single stuck-at faults with a test program consisting of 1,129 bytes, running for a total of 137,649 clock cycles.

In 2001, the problem of processor testing and processor-based SoC testing was addressed by W.-C.Lai and K.-T.Cheng [104], where instruction-level DfT modifications to the embedded processor were introduced. Special instructions are added to the processor instruction set with the aim to reduce the length of a self-test program for the processor itself or for other cores in the SoC, and to increase the obtained fault coverage. Experimental results on two simple processor models, the Parwan [116] and the DLX [59], show that complete coverage of all functionally testable path delay faults can be obtained with small area DfT overheads that also reduce the overall self-test program length and its total execution time. In Parwan, test program is reduced by 34% and its execution time reduced by 39% with an area overhead of 4.7% compared to the case when no instruction-level DfT is applied to the same processor [105]. Moreover, complete 100% fault coverage is obtained while in [105] fault coverage was a bit lower 99.8%. In the DLX case, the self-test program is 15% smaller and its execution time is reduced by 21% with an area overhead due to DfT of only 1.6%. Fault coverage is complete (100%) for the DfT case while it was 96.3% in the design without the DfT modifications [105]. All fault coverage numbers refer to the set of functionally testable path delay faults.

In 2001, F.Corno, M.Sonza Reorda, G.Squillero and M.Violante [32] presented a functional testing approach for microprocessors. The approach consists of two steps. In the first step, the instruction set of the processor is used for the construction of a set of macros for each instruction. Macros are responsible for the correct application of an instruction and the observation of its results. In the second step, a search algorithm is used to select a suitable set of the previously developed macros to achieve acceptable fault coverage with the use of ATPG generated test patterns. A genetic algorithm is employed in the macros selection process to define the values for each of the parameters of the macros. Experimental results are reported in [32] on the 8051 microcontroller. The synthesized circuits consist of about 6,000 gates and 85.19% single stuck-at fault coverage is obtained, compared to 80.19% of a pure random-based application of the approach. The number of macros that actually contributed to the above low fault coverage, lead to a test program consisting of 624 processor instructions.

In 2002, L.Chen and S.Dey exploited the fault diagnosis capability of software-based self-testing [29]. A large number of appropriately developed test programs are applied to the processor core in order to partition the fault universe in smaller partitions with unique pass/fail pattern. A sufficient diagnostic resolution and quality was obtained when the approach is applied to the simple educational processor Parwan [116].

In 2002, N.Kranitis, D.Gizopoulos, A.Paschalis and Y.Zorian [94], [95], [96] introduced an instruction-based self-testing approach for embedded processors. The self-test programs are based on small deterministic sets of test patterns. First experimental results for the methodology were presented in these papers. The approach was applied to Parwan, the same small accumulator-based processor used in [28] and a 91.34% single stuck-at faults coverage was obtained with a self-test program consisting of around 900 bytes and executing for about 16,000 clock cycles (these numbers gave about 20% reduction of test program size and about 90% reduction in test execution time compared to [28]).

In 2002, P.Parvathala, K.Maneparambil and W.Lindsay presented an approach called Functional Random Instruction Testing at Speed (FRITS) which applies randomized instruction sequences and tries to reduce the cost of functional testing of microprocessors [125]. To this aim DfT modifications are proposed that enable the application of the functional self-testing methodology using low-cost, low-pin count testers. Moreover, automation of the self-testing programs generation is considered. The basic feature of FRITS which is also its main difference when compared with classical functional processor self-testing is that a set of basic FRITS routines (called kernels) are loaded to the cache memory of the processor and are responsible for the generation of several programs consisting of random instruction sequences and are used to test parts of the processor. External memory cycles are avoided by appropriate exception handling that eliminates the possibility of cache misses that initiate main memory accesses. The FRITS methodology is reported to be applied in the Intel Pentium® 4 processor resulting in around 70% of single stuck-at fault coverage. Also, application of the approach to the Intel Itanium™ processor integer and floating point units led to 85% single stuck-at fault coverage. The primary limitation of this technique, like any other random-based, functional self-testing technique, is that an acceptable level of fault coverage can only be reached if very long instruction sequences are applied – this is particularly true for complex processor cores with many components.

In 2003, L.Chen, S.Ravi, A.Raghunathan and S.Dey focused on the scalability and automation of software-based self-testing [31]. The approach employs RTL simulation-based techniques for appropriate ranking and selection of self-test program templates (instruction sequences for test delivery to each of the processor components) as well as techniques from the theory of statistical regression for the extraction of the constraints imposed to the application of component-level tests by the instruction set of the processor. The constraints are modeled as virtual constraint circuits (VCC) and automated self-test program generation at the component level is performed. The approach has been applied to a relatively large combinational sub-circuit of the commercial configurable and extensible RISC processor Xtensa™ from Tensilica [174]. A self-test program of 20,373 bytes, running for a total of 27,248 clock cycles obtained 95.2% fault coverage of the functionally testable faults of the component. A total of 288 test patterns generated by an ATPG were applied to the component.

In 2003, N.Kranitis, G.Xenoulis, D.Gizopoulos, A.Paschalis and Y.Zorian [99], [100], showed the applicability of software-based self-testing to larger RISC processor models while a classification scheme for the different processor components was introduced. The processor self-testing problem is addressed by a solution focusing on the test costs reduction in terms of engineering effort, self-test code size and self-test execution time, all together leading to significant reductions of the total device test time and costs. In particular, the work of [100] gives an extensive application of the low-cost, deterministic self-testing methodology to several different implementations of the same RISC processor architecture. The processor model consists of 26,000 to 30,000 logic gates depending on the implementation library and synthesis parameters. A self-test program of 850 instructions (words) executing in 5,800 clock cycles obtained more than 95% single stuck-at fault coverage for all different processor implementations.

In 2003, G.Xenoulis, D.Gizopoulos, N.Kranitis and A.Paschalis discussed the use of software-based self-testing as a low-cost methodology for on-line periodic testing of processors [173]. The basic requirements of such an approach were discussed and the coding styles for software-based self-testing that are able to satisfy the requirements were presented.

4.2.2 Industrial Microprocessors Testing

Apart from the research described in the previous sections which focused on more or less generic methodologies for processor testing and self-testing, a large set of industrial case studies papers were presented at the IEEE International Test Conference (ITC) over the last years. In this type of papers, authors of major microprocessor design and manufacturing companies summarized the manufacturing testing methodologies applied to several generations of successful, high-end microprocessors of our days. Several exciting ideas were presented in these papers although, for obvious reasons, only a small set of the details were revealed to the readers.

A list of papers of this category is included in the bibliography of this book including microprocessor testing papers from Intel, IBM, Motorola, Sun, HP, AMD, DEC, ARM, TI. The list includes industrial papers presented at ITC between years 1990 and 2003.

4.3 Classification of the Processor Testing Methodologies

Each of the techniques briefly discussed in the previous sections belongs to one or more of the categories discussed in the beginning of the Chapter. In this section we make the association of each methodology with the processor testing categories it belongs to. This association is sometimes tricky because a methodology originally developed for one objective may also be applicable with other objectives in mind and in a different application field. Therefore, the main intention of the classification of this section is to give a rough idea of the work performed over the years with emphasis on several aspects of processor testing.

Table 4-8 presents the classification of the works analyzed above into the categories described earlier. Some explanations are necessary for the better understanding of Table 4-8.

A paper appearing in the self-testing category means that it specifically focused on self-testing but in many cases the approach can be applied to external testing as well.

A paper appearing in the DfT-based testing category means that some design changes are required to the processor structure while all other works do not change the processor design.

A paper appearing in the sequential faults testing category means that the approach was developed and/or applied targeting sequential faults such as delay faults. All other works were either applied for combinational faults testing or did not use any structural fault model at all (functional testing).

Category	Works
Self Testing *(others are on external testing only)*	[17], [161], [21], [86], [47], [91], [18], [2], [131], [175], [10], [105], [28], [27], [104], [94], [96], [95], [125], [31], [97], [100], [99]
DfT-based Testing *(others are non-intrusive)*	[77], [2], [104]
Functional Testing	[135], [156], [158], [157], [34], [137], [139], [149], [6], [77], [21], [45], [86], [47], [64], [141], [142], [12], [91], [146], [145], [152], [119], [58], [109], [108], [165], [166], [144], [32], [125]
Structural Testing *(including register transfer level)*	[149], [143], [146], [145], [2], [131], [61], [175], [105], [28], [27], [104], [29], [94], [95], [96], [31], [99], [97], [100]
Sequential Faults Testing *(others are on combinational faults)*	[105], [104]
Pseudorandom Testing	[158], [157], [159], [161], [160], [41], [12], [91], [148], [131], [144], [175], [10], [32], [125]
Deterministic Testing *(including ATPG-based)*	[148], [28], [27], [94], [96], [95], [97], [100], [99], [166], [165], [105], [28], [27], [32], [31]
Diagnosis *(others are on testing only)*	[137], [138], [140], [168], [29]
Field Testing – On-Line Testing *(others are on manufacturing testing)*	[33], [92], [64], [107], [135], [173]
DSP Testing *(others are on microprocessor testing)*	[152], [2], [131], [175]

Table 4-8: Processor testing methodologies classification.

Chapter 5

Software-Based Processor Self-Testing

In this Chapter we discuss a processor self-testing approach which has recently attracted the interest of several test technologists. The approach is based on the execution of embedded self-test programs and is known as *software-based self-testing*. We present software-based self-testing as a low-cost or cost-effective self-testing technique that aims to high structural fault coverage of the processor at a minimum test cost.

The principles of software-based self-testing and the methodology outline are presented in this Chapter in a generic way that does not depend on any particular processor's architecture. Different alternatives in the application of software-based self-testing are also discussed.

Through a detailed discussion of the idea, its requirements and objectives and its actual implementation in several practical examples of embedded processors that follows in the subsequent Chapter, it is shown that software-based self-testing, if appropriately realized and carefully tailored for the needs of a particular processor's design, can be an effective and efficient strategy for low-cost self-testing of processor cores.

Low-cost processor testing has several different facets: it may refer to the test generation time and man effort dedicated to the development of a self-testing strategy for a processor core; it may also refer to the total test application time for the processor chip, while it may also refer to the extra hardware dedicated to self-testing, and the performance overhead or the power consumption overhead of a particular self-testing flow. We elaborate

Embedded Processor-Based Self-Test
D.Gizopoulos, A.Paschalis, Y.Zorian

on these aspects of test cost in the context of software-based self-testing for embedded processors.

In this Chapter, we present the steps of the software-based self-testing philosophy, each one of them with different alternatives, depending on the processor architecture and the application requirements. The presentation style is generic, and different approaches found in the literature (already analyzed in Chapter 4) can be selectively applied as parts of the methodology.

The analysis of software-based self-testing given in this chapter is followed, in the next Chapter, by a comprehensive set of experimental results on several different publicly available processor architectures. Different Instruction Set Architectures (ISA) are studied and the effectiveness of software-based self-testing on each of them is discussed.

5.1 Software-based self-testing concept and flow

The concept of software-based self-testing has been introduced in the previous Chapters, but for completeness of this Chapter, we present again the basic idea and we subsequently proceed to the details and the steps of its application. Figure 5-1 depicts the basic concept of processor testing by executing embedded self-test software routines.

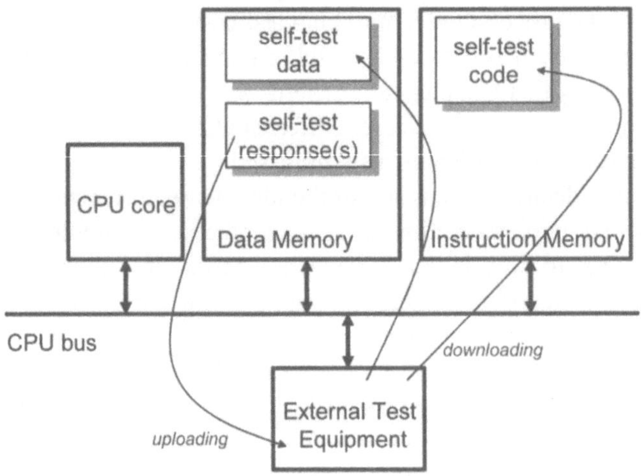

Figure 5-1: Software-based self-testing for a processor (manufacturing).

Application of software-based self-testing to a processor core's manufacturing testing consists of the following steps:

- The self-test code is downloaded to the embedded instruction memory of the processor via external test equipment which has access to the internal bus[14]. The embedded code will perform the self-testing of the processor. Alternatively, the self-test code may be "built-in" in the sense that it is permanently stored in the chip in a ROM or flash memory (this scenario is shown in Figure 5-2). In this case, there is no need for a downloading process and the self-test code can be used many times for periodic/on-line testing of the processor in the field.

- The self-test data is downloaded to the embedded data memory of the processor via the same external equipment. Self-test data may consist, among others, of: (i) parameters, variable and constants of the embedded code, (ii) test patterns that will be explicitly applied to internal processor modules for their testing, (iii) the expected fault-free test responses to be compared with actual test responses. Downloading of self-test data does not exist if on-line testing is applied and the self-test program is permanently stored in the chip.
- Control is transferred to self-test program which starts execution. Test patterns are applied to internal processor components via processor instructions to detect their faults. Components' responses are collected in registers and/or data memory locations. Responses may be collected in an unrolled manner in the data memory or may be compacted using any known test response compaction algorithm. In the former case, more data memory is required and test application time may be longer, but, on the other side, aliasing problems are avoided. In the latter case, data memory requirements are smaller because only one, or just a few, self-test signatures are collected, but aliasing problems may appear due to compaction.
- After self-test code completes execution, the test responses previously collected in data memory, either as individual responses for each test pattern or as compacted signatures are transferred to the external test equipment for evaluation.

In the discussion of software-based self-testing of this Chapter, we mainly focus on manufacturing testing and in several locations we point out the differences when software-based self-testing is used for periodic, on-line testing in the field. In the case of periodic, on-line testing there is no need to transfer self-test code, data and responses to and from external test equipment (see Figure 5-2). Self-test code, data and expected response(s) (compacted or not) are stored in the chip as part of the design (in a ROM for example). Execution of self-test programs leads to a pass/fail indication

[14] In general, it is assumed that a mechanism exists for the transfer of self-test code and data to the embedded instruction and data memory. This mechanism can be for example a simple serial interface or a fast Direct Memory Access (DMA) protocol.

which can be used subsequently for further actions on the system (repair, re-configuration, re-computation, etc).

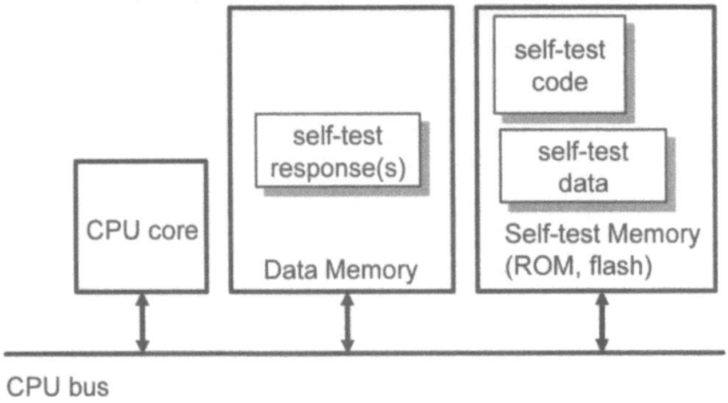

Figure 5-2: Software-based self-testing for a processor (periodic).

The steps of software-based self-testing described in the list above can be applied in a different way depending on the processor-based system architecture and configuration; we elaborate on this in the rest of the Chapter. Two factors that have significant impact on the actual implementation details of software-based self-testing are the *test quality* that is looked for and the *test cost* that can be afforded in a particular system. These factors determine the permissible limits for the self-test code size and execution time as well as other details of software-based self-testing.

Let us study in more detail, at the processor level, how test patterns are applied to internal processor components by the execution of processor instructions, how faults in these components are excited and how their effects are propagated outside the processor core for observation and eventually for fault detection. We present this process in a generic way that does not depend on any particular processor's instruction set.

Application of test patterns to processor component via processor instructions consists of the following three steps:

- *Test preparation:* test patterns are placed in locations (usually registers but also in memory locations) from which they can be easily applied to a processor component (the component under test). This step may require the execution of more than one processor instructions[15].
- *Test application and response collection:* test patterns are applied to the processor's component under test and the component's response(s) are collected in locations (usually registers but may

[15] For example, a test pattern for a two-operand operation consists of two parts, the first and the second operand value. Application of such a test pattern requires two register writes.

also be memory locations). This step usually takes one instruction.

- *Response extraction:* responses collected internally are exported towards data memory (if not already in memory by the test application instruction). This step may also require the execution of more than one instruction[16].

We note that the above steps can be partially merged depending on the particular coding style used for a processor and also the details of its instruction set. For example, if the processor instruction set contains instructions that directly export the results of an operation (ALU operation, etc) to a memory location then the steps of response collection and response extraction are actually merged. Figure 5-3 shows in a graphical way the three steps of software-based self-testing application to an internal component of a processor core. All three steps are executed by processor instructions; no additional hardware has to be synthesized for self-testing since the application of self-testing is performed by processor instructions that use existing processor resources. The processor is not placed in a special mode of operation but continues its normal operation executing machine code instructions.

[16] This is the case when the test application instruction gives a multi-word result (two-word results are given by multiplication and division), or the extraction of a data word and a status word (a status register) is necessary.

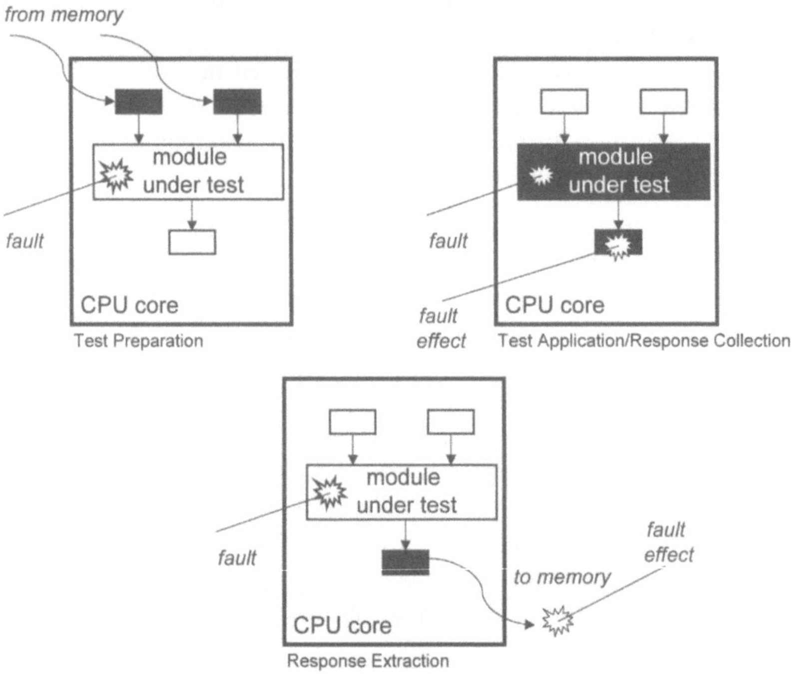

Figure 5-3: Application of software-based self-testing: the three steps.

The process of applying a test pattern to a processor component using its instruction set can be further clarified if we have a look at a very simple example using assembly language pseudocode which is very much alike the assembly language of most modern RISC processors (a classic RISC load-store architecture). The following assembly language pseudocode portion shows the steps of Figure 5-3 in the case that the component under test is an arithmetic logic unit (ALU) and the instruction applied to it is an addition instruction. The two operands of the operation come from two processor registers while the result of the addition is stored to a third processor register.

```
load r1, X        -- test preparation step 1
load r2, Y        -- test preparation step 2
add r3, r1, r2    -- test application/response collection
store r3, Z       -- response extraction
```

The registers used in the example are named **r1**, **r2**, **r3**, while **X**, **Y** are memory locations each of which contain half of the test patterns to be applied to the ALU and **Z** is the memory location where the test response of the ALU is eventually stored[17]. The test preparation step consists of loading the two registers **r1**, **r2** with part of the test pattern to be applied to the ALU. Memory locations **X** and **Y** contain these two parts of the ALU test pattern. Registers **r1** and **r2** (like all general purpose registers of the processor) have access to the ALU component inputs. The test application and response collection step consists of the execution of the addition instruction where the test pattern now stored in **r1** and **r2** is applied to the ALU and its response (result of the addition) is collected into register **r3** (in a single instruction **add**). Finally, the **store** instruction is the response extraction step where the ALU response which is stored in register **r3** is transferred from inside of the processor to the outside external memory location **Z**. As we have already mentioned, test responses may either be collected in memory locations individually, or compacted into a single or a few signatures to reduce the memory requirements.

Of course, not every module of the processor can be tested with such simple code portions as the one above, but this assembly language code shows the basic simple idea of software-based self-testing. The effectiveness of software-based self-testing depends on the way that test generation (i.e. generation of the self-test programs) is performed to meet the specific requirements of an application (test code size, test code execution time, etc). Moreover, it strongly depends on the Instruction Set Architecture of the processor under test. The requirements for an effective application of software-based self-testing are discussed in the following subsection.

5.2 Software-based self-testing requirements

Software-based self-testing is by its definition an effective processor self-testing methodology. Its basic objective is to be an alternative self-testing technique that eliminates or reduces several bottlenecks of other techniques.

Specifically, software-based self-testing is effective because:

- it does not add hardware overhead to the processor but rather it uses existing processor resources (functional units, registers, control units, etc) to test the processor during its normal mode of operation executing embedded code routines;
- it does not add performance overhead and does not degrade the circuit normal operation because nothing is changed in its well-optimized structure and critical paths remain unaffected;

[17] In this example, the test patterns are stored in locations of the data memory (words X and Y). As we will see, test patterns can be also stored in the instruction memory, as parts of instructions when the immediate addressing mode is used in self-test programs.

- it does not add extra power dissipation during the processor's normal operation;
- it does not add extra power dissipation when the self-test programs are executed, simply because self-test programs are just like normal programs executed when the processor performs its normal operation[18];
- it does not rely on expensive external testers, but rather on low-cost, low pin-count, testers which are only necessary for the transfer of the self-test programs (code and data) to the processor's memory and for the transfer of the self-test responses out of the processor's memory for external evaluation.

Software-based self-testing applies tests that can appear during the processor's normal operation. Therefore, the faults that it can detect are the functionally testable faults of the processor (either the logic faults or the timing faults). This property of software-based self-testing has a positive and a negative impact. On the positive side, no overtesting is applied, i.e. only faults that could affect the normal operation of the processor are targeted. On the negative side, there is no mechanism (no DfT circuitry) that increases the testability of the processor in its hard-to-test areas.

The actual efficiency of software-based self-testing on specific processor architectures depends on system parameters such as the total system cost (and thus its test cost) and the required manufacturing testing quality. A number of questions must be answered before a test engineer applies software-based self-testing to a processor. The answers to these questions construct a framework for the development of self-test programs. In the next sections, we elaborate on these subjects as well as on their impact on the success of software-based self-testing.

5.2.1 Fault coverage and test quality

The primary and most important decision that has to be taken in any test development process is the level of fault coverage and test quality that is required. All aspects of the test generation and test application processes are affected by the intended fault coverage level: is 90%, 95% or 99% fault coverage for single stuck-at faults sufficient?

Another issue related to test quality is the fault model that is used for test development. Comprehensive sequential fault models, such as the delay fault

[18] Power consumption concerns during manufacturing testing are basically related with the stressing of the chip's package and thermal issues. Power consumption concerns during on-line periodic testing are related with the duration of the system's battery which may be exhausted faster if long self-test programs are executed.

model, are known to lead to higher test quality and higher defect coverage than traditional, simpler combinational fault models such as the stuck-at fault model. On the other hand, sequential fault models consume significantly more test development time and also lead to much larger test application time because of the larger size of the test sets they apply to the circuit.

In summary, in software-based self-testing a hunt for higher fault coverage and/or a more comprehensive fault model under consideration means:

- more test engineering effort and more CPU time for manual and automatic (ATPG-based) self-test program generation, respectively;
- larger self-test programs and thus longer downloading time from external test equipment to the processor's memory;
- longer test application time, i.e. longer self-test program execution intervals, thus more time that each chip spends during testing.

When a self-test program guarantees very high fault coverage for a comprehensive fault model, this denotes a high quality test strategy.

In software-based self-testing, there is also a fundamental question that must be answered by any potential methodology: what extent of fault coverage and test quality *is at all feasible*?

Software-based self-testing being a non-intrusive approach may not be able to achieve fault coverage levels that test approaches based on structured DfT techniques can obtain. Software-based self-testing is capable to detect faults that can possibly appear in normal operation of the circuit and therefore performs only the absolutely necessary testing to the chip. This way the problem of *overtesting* the processor is avoided. Overtesting can happen when a chip is rejected as faulty as a consequence of the detection of faults that can never happen while the chip operates in its normal mode. In structured DfT based testing techniques like scan-based testing, chips are tested under non-functional mode of operation and therefore overtesting can happen with severe impact on production yield. On the other side, the existence of DfT infrastructure in a chip makes its testing easier.

A basic issue throughout this Chapter is the total test cost of software-based self-testing. Therefore, in subsequent sections, we will try to quantify the test cost terms and thus explain the reasons why software-based self-testing is a low-cost test methodology.

5.2.2 Test engineering effort for self-test generation

A major concern in test generation for electronic systems is the man-power and related costs that are spent for test development. This cost factor is an increasingly important one as the complexity of modern chips increases. In processors and processor-based designs, software-based self-testing provides a low-cost test solution in terms of test development cost.

Any software-based self-testing technique can theoretically reach the maximum possible test quality in terms of fault coverage (i.e. detection of all faults of the target fault model that can appear during system's normal operation) if an unlimited time can be spent for test development. Of course, unlimited test development time is never allowed! A limited test engineering time and corresponding labor costs can be dedicated during chip development for the generation of self-test programs.

The ultimate target of a testing methodology is summarized in the following sentence: *obtain the maximum possible fault coverage/test quality under specific test development and test application cost constraints.* Therefore, if a methodology is capable to reach a high fault coverage level with small test engineering effort in short time, it is indeed a cost effective test solution and particularly useful for low-cost applications. This aspect of software-based self-testing is the one that is analyzed more in this Chapter.

Of course, if a particular application's cost analysis allows an unlimited (or very large) effort or time to be spent for test development, then higher fault coverage levels can be always obtained.

When discussing software-based self-test generation in subsequent sections we focus on the importance of different parts of the processor for self-test programs generation so that high fault coverage is obtained by software self-test routines *as quickly as possible.* By applying this "greedy", in some sense, approach, limited test engineering effort is devoted to test generation to get the maximum test quality under this restriction.

Figure 5-4 presents in a graphic way the main objective of software-based self-testing as a low-cost testing methodology. Between Approach A and Approach B both of which may eventually be able to reach a high fault coverage of 99%, the best one is Approach A since it is able to obtain a 95%+ fault coverage quicker than Approach B. "Quicker" in this context means with less engineering effort (man power) and/or with less test development costs. This is basic objective of software-based self-testing as we describe it in this Chapter.

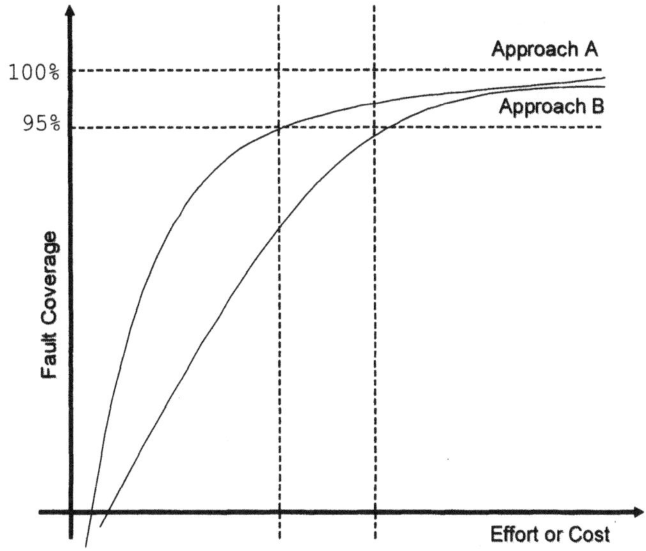

Figure 5-4: Engineering effort (or cost) versus fault coverage.

When devising a low-cost test strategy, the most important first step is to identify the cost factor or factors, if more than one, that are the most important for the specific test development flow. Low test engineering effort may be one of the important cost factors because it is calculated in personnel cost for test development (either manual or EDA assisted). But, on the other side, test development is always a one-time effort that results to a self-test program which is eventually applied to all the same processors. Therefore, the test development cost and corresponding test engineering efforts are divided by the number of devices that are finally produced. In high volume production lines test development costs have marginal contribution to the overall system costs. If production volume is not very high then test development costs should be more carefully considered in the global picture. This is the case of low-cost applications.

5.2.3 Test application time

Total test application time for each device of a production line has a direct relation with the total chip manufacturing cost. In particular, test application time in software-based self-testing consists of three parts:

- self-test program download time from external test equipment into embedded memory;
- self-test program execution and responses collection time;
- self-test responses upload time from embedded memory to external test equipment for evaluation

The relation between the frequency of the external test equipment and the frequency of the processor under test determines the percentage of the total test application time that belongs to the downloading/uploading phases or the self-test execution phase. The basic objective of software-based self-testing is to be used as a low-cost test approach that utilizes low-cost, low-memory and low pin-count external test equipment. To this aim the most important factors that must be optimized in test application is the time of the downloading of the test program and the uploading of the test responses, i.e. the first and third parts of software-based self-testing as given above. A simple analysis of the total test application time of software-based self-testing is useful to reveal the impact of all three parts of test application time.

Let us consider that the processor under test has a maximum operating frequency of f_{uP} while the external tester[19] has an operating frequency of f_{tester}. This means that a self-test program (code and data) can be downloaded at maximum rate of f_{tester} and it can be executed by the processor at a maximum rate of f_{uP}.

There are two extremes for the downloading of the self-test program into the embedded memory of the processor. One extreme is the parallel downloading of one self-test program word (instruction on data) per cycle. Fast downloading of self-test programs can be also assisted if the chip contains a fast Direct Memory Access (DMA) mechanism which is able to transfer large amounts of memory contents without the participation of the processor in this transfer. The other extreme is the serial downloading of a self-test program bit per cycle. This is applicable is case that a serial interface is only available in the chip for self-test program downloading but this is a rare situation. If the total size of self-test program (code and data) is small enough, then even the simple serial transfer interface will not severely impact the total self-test application time.

Let us also consider that a self-test program consisting of C instruction words and D data words must be downloaded to the processor memory (instruction and data memory) and that a final number of R responses must be eventually uploaded from processor data memory to the tester memory for external evaluation. Finally, we assume that the execution of the entire self-test program and the collection of all responses take in total K clock cycles of the processor. For simplicity, we assume that the K clock cycles include any stall cycles that may happen during self-test program execution either for memory accesses (memory stall cycles) or for processor internal reasons (pipeline stall cycles).

[19] We use the term "tester" denoting any external test-assisting equipment. It can be an expensive high-frequency tester, a less expensive tester or even a simple personal computer, depending on the specific application and design.

The total test application time for software-based self-testing is roughly given by the following formula:

$$T = (C+D)/f_{tester} + K/f_{uP} + R/f_{tester}, \quad \textit{or}$$

$$T = (C+D+R)/f_{tester} + K/f_{uP}, \quad \textit{or}$$

$$\mathbf{T = W/f_{tester} + K/f_{uP}}, \quad \textit{where } \mathbf{W = C+D+R}$$

It is exactly this last formula that guides the process of test generation for software-based self-testing and the embedded software coding styles, as we will see in subsequent sections. The relation between the f_{tester} and f_{uP} frequencies reveals the upper limits for the self-test program size (instructions – C, and data – D) and responses size (R) as well as for the self-test program execution time (number of cycles K).

Figure 5-5 and Figure 5-6 use the above formula to show in a graphical way, how the total test application time (T) of software-based self-testing is affected by the relation between the frequency of the chip (f_{uP}) and frequency of the tester (f_{tester}) as well as by the relation between the number of clock cycles of the program (K) and its total size (W).

Figure 5-5 presents the application time of software-based self-testing as a function of the ratio K/W, for three different values of the ratio f_{uP}/f_{tester}. A large value of the K/W ratio means that the self-test program (instructions+data+responses) is small and compact (smaller value for W) and/or it consists of loops of instructions executed many times each (small code but large execution time). Obviously as we see in Figure 5-5, in all three cases of f_{uP}/f_{tester} ratios (2, 4 and 8), when the K/W ratio increases the total test application time increases. The most important observation in the plot is that when the f_{uP}/f_{tester} ratio is smaller (2 for example), this test application time increase is much faster compared to larger f_{uP}/f_{tester} ratios (4 or 8). In other words, when the chip is much faster than the tester then an increase in the number of clock cycles for self-test execution (K) has a small effect on the chip's total test application time. On the other side, if the chip is not very fast compared to the tester, then an increase in the number of clock cycles leads to a significant increase in test application time.

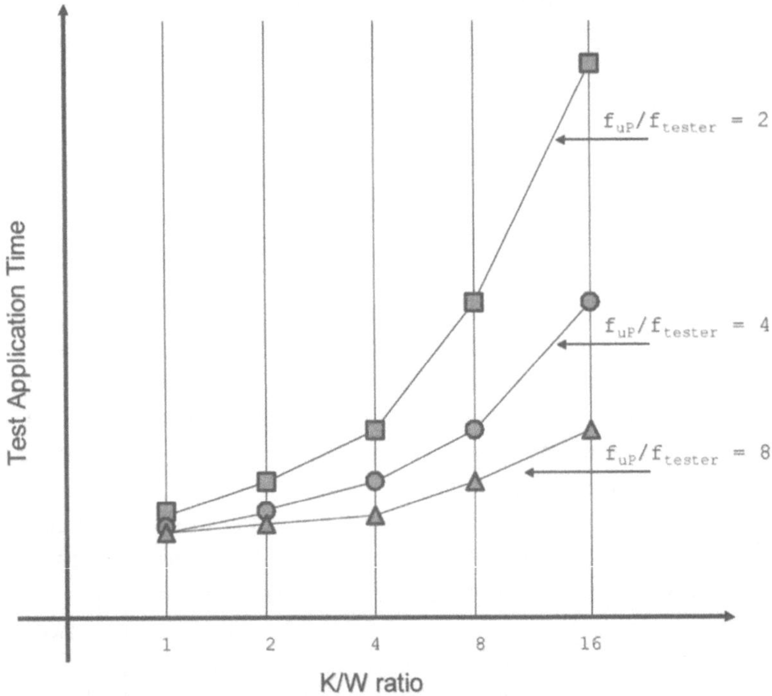

Figure 5-5: Test application time as a function of the K/W ratio.

From the inverse point of view, Figure 5-6 shows the application time of software-based self-testing as a function of the ratio f_{uP}/f_{tester}, for three different values of the ratio K/W. An increasing value of the f_{uP}/f_{tester} ratio denotes that the processor chip is much faster than the tester. This means that for the same ratio K/W the test application time will be smaller since the program will be executed faster. This reduction becomes sharper and more significant when the K/W ratio has smaller values (2 for example), i.e. when the self-test program is not so compact in nature and the number of clock cycles is close to the size of the program.

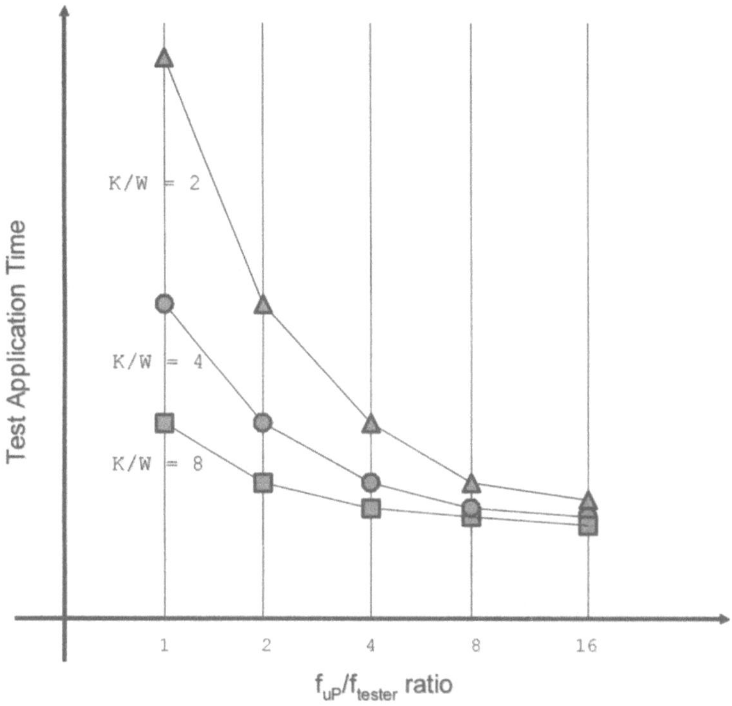

Figure 5-6: Test application time as a function of the f_{uP}/f_{tester} ratio.

In the above discussion, W is the sum of the self-test code, self-test data and self-test responses, C, D and R, respectively. When test responses of the processor components are compacted by the self-test program (a special compaction routine is part of the self-test program) then R is a small number, possibly 1, when a single signature is eventually calculated. This leads to a reduction of time for uploading the test responses (signatures) but on the other side, the self-test program execution time (number of clock cycles K) is increased because of the execution of the compaction routines.

We have to point out that the above calculations and values of Figure 5-5 and Figure 5-6 are simple and rough approximations of the test application time of software-based self-testing based only on the time required to download/upload the self-test program (code, data and responses) and the time required to execute the self-test program and collect the test responses from the processor. A more detailed analysis is necessary for each particular application, but the overall conclusions drawn above will still be valid for the impact of the f_{uP}/f_{tester} and K/W ratios on the test application time of software-based self-testing.

In the case of on-line periodic testing of the processor, the download/upload phases either do not exist at all or are executed only at system start-up[20]. Therefore, the self-test program size does not have an impact on the test application time in on-line testing. On the other side, the size of the self-test program is important in on-line periodic testing because it is related with the size of the memory unit where it will be stored for periodic execution.

5.2.4 A new self-testing efficiency measure

In external testing or hardware-based self-testing, the number of applied test patterns that are necessary to attain certain fault coverage gives a measure of the test efficiency. The number of test patterns multiplied by the test application frequency (one test pattern per clock cycle or one test pattern per scan) determines the overall test application time.

In software-based self-testing the number of applied test patterns do not directly correspond to the overall test application time because this depends on the number of clock cycles than a self-test program needs before each test pattern is applied. Additionally, the test application time for (manufacturing) software-based self-testing has another significant portion: the self-test program downloading time.

Based on the brief analysis for the test application time of software-based self-testing given in the previous section 5.2.3, we define a new measure for the efficiency of software-based self-testing that includes all parameters that determine test application time.

The new measure is called *sbst-duration (SD)* and is defined as:

$$SD = W + K \times Q$$

where, $W=C+D+R$ and K are as defined in section 5.2.3 (number of words for code, data and responses, respectively) and $Q=f_{tester}/f_{uP}$ (the ration between the tester and the processor frequencies).

The *sbst-duration* measure gives a simple, frequency-independent idea of the duration of software-based self-testing, just like the number of test patterns do for external testing or hardware-based self-testing. The actual test application time can be obtained by multiplying sbst-duration with the external tester's frequency.

Using the *sbst-duration* measure, different software-based self-testing methodologies can be compared independently of the actual frequencies of the processor or the external test equipment. The same fault coverage for the processor can be possibly by self-test programs with different sbst-duration

[20] If the self-test code and data are stored in a memory unit such as a ROM, then there is no download/upload phase. If they are transferred from a disk device into the processor RAM, this will only happen once at system start-up.

measures. The one with the smallest measure can be easily considered more efficient at a given ration between the frequencies of the external tester and the processor.

5.2.5 Embedded memory size for self-test execution

Embedded software routines that are executed in a processor are fetched from embedded memory that stores the instruction sequences as well as their data. In software-based self-testing, the maximum size of a self-test program (code and data) may be restricted by the size of available on-chip memory.

In the case of manufacturing testing of processors using software-based self-testing, the memory used for self-test program execution is the on-chip cache memory. Self-test programs in this case must be designed so that external memory cycles are eliminated (caused by cache misses). In a SoC application where limited cache memory is available on-chip, software self-testing routines should not exceed in size the memory size. If software self-testing routines can't be accommodated in the cache memory entirely, then multiple loadings of the memory with portions of the self-test code and data are necessary with a corresponding impact on the total test application time of the chip. If self-test code and data are stored in main memory instead of cache memory, then the available memory size is larger but the self-test execution time will be longer.

On-chip memory limitations are more critical when software-based self-testing is used for on-line periodic testing while the system operates in its normal environment. In this case, software self-testing programs are regular "competitors" of system resources with normal programs executed by the user and thus available instruction and data memory is usually much less than during manufacturing testing where the entire memory is available can be used for software-based self-test execution.

In on-line periodic testing, self-test code and data may be permanently stored in the system, so that it can be executed periodically (during idle intervals or in intervals where the system normal execution is paused). Therefore, the size of the self-test program and data is a serious cost factor for the system in terms of extra silicon area occupied by the dedicated memory (a ROM or flash memory are both appropriate candidates for such a configuration). If a permanently stored self-test program leads to serious cost increase, then an alternative way to apply on-line, software-based self-testing is to load the self-test program to the chip's memory units at system's start-up. This happens once and self-test program is available for periodic execution but it permanently occupies part of the system's memory.

5.2.6 Knowledge of processor architecture

The effectiveness and efficiency of any test development strategy is related to the level of information that is necessary to be available for the circuit under test. In the case of processor cores, if only one piece of information is available this must be the processor's Instruction Set Architecture, which includes:

- the processor's instruction set: instruction types, assembly language syntax and operation of the machine language instructions;
- the processor's registers which are visible to the assembly language programmer;
- the addressing modes for operands reference used by the processor's instructions;

The Instruction Set Architecture information is always available for a processor in the programmer's manual and can be considered a minimum basis for software-based self-test generation. As we mentioned in Chapter 4, functional testing methodologies for processors can be based only on this information.

At the other end of the spectrum, structural testing techniques (including structural software-based self-testing techniques) for processors, require availability of the low-level details of the processor. This means that gate level implementation details of the processor components are necessary for test generation[21]. If such low-level information for the processor under test actually exists and is available to the test engineer and EDA tools for (possibly constraints-based) test generation, then this flow can lead to very high fault coverage with limited engineering effort. In many cases, gate level information of an embedded processor is not available for automatic test generation at the structural level, but even if it was, automatic test pattern generation for complex sequential machines like processor can't be handled even by the most sophisticated EDA tools[22].

The objective of software-based self-testing, as a low-cost self-testing methodology, is to be able to reach relatively high fault coverage and test quality levels at the expense of small test engineering effort and cost. In order to preserve the generality of the approach and make it attractive to

[21] Either a gate-level netlist must be available or a synthesizable model of the processor from which a gate-level netlist can be generated after synthesis.

[22] Even non-pipelined processors are very deep sequential circuits, and sequential ATPG tools usually fail to generate sufficiently small test sets that reach high fault coverage for a processor.

most applications, as few as possible information about the processor's architecture and implementation should be required.

Software-based self-testing is generic processor testing methodology which can be used independently of the level of information know for the processor core (soft core, firm core or hard core). Its efficiency may vary from case to case depending on the available information. The more structural information is available the higher fault coverage can be obtained.

5.2.7 Component based self-test code development

Like every divide-and-conquer approach, if *component-based* self-test routines development is used in software-based self-testing it is able to manage the complexity of test generation for embedded processors. Processors are sophisticated sequential machines where all known design methods are usually applied to get the best performance of the circuit. Even the most advanced ATPG tools are not able to develop sufficient sets of test patterns for such complex architectures. Dividing the problem of processor test generation and in particular self-test routines development into smaller problems for each of the components, makes the solution of the problem feasible.

Component-based test development, allows a selective approach where the most important components are targeted for test generation, so that the highest possible structural fault coverage is obtained with the smallest test development effort and cost. As we will see in subsequent sections of this Chapter, the most important components of the processor in terms of contribution to the total fault coverage are considered first.

If the test generation problem is considered at the top-level of the processor hierarchy (the entire processor) without using ATPG tools but following a pseudorandom philosophy (use of pseudorandom instruction sequences, pseudorandom operations and pseudorandom operands), then serious drawbacks, such as low fault coverage and long test application time, have been extensively reported in the literature. These drawbacks are due to the fact that processors are inherently pseudorandom resistant circuits.

A last but not least indication that component-based self-test routines development is indeed an effective approach to software-based self-testing, is that the most successful works in the literature (see Chapter 4) that report successful application and relatively high structural fault coverage results on software-based self-testing for processors are those that are component-based in nature.

5.3 Software-based self-test methodology overview

In this section we give an overview of software-based self-testing and its breakdown into different phases. Subsequently, we analyze each of the phases along with simple informative examples.

As we have already stated, the intention of this book is not to describe a single approach for software-based self-testing of processor cores, but on the contrary to discuss the overview and specific details of software-based self-testing with the final aim to make clear that it is a *low-cost self-test* solution that targets high structural fault coverage for processors.

Although several *functional* self-testing techniques have been applied in the past to processors (see the bibliography review given in Chapter 4), they are not always suitable for low-cost self-testing because of two main reasons:

- they don't target high fault coverage for a particular structural fault model but rather they focus on coverage of a functional fault model or a function-related metric, therefore, they don't focus on structural testing and fail to obtain high structural fault coverage;
- they are mostly pseudorandom-based and rely on the application of pseudorandom instruction sequences, pseudorandom operations, pseudorandom operands or a combination of these three; due to this pseudorandom nature functional testing approaches for processors require very long instruction sequences and/or very large program sizes for self-testing, while at the same time, they are unable to reach the high levels of structural fault coverage because of the random pattern resistance of several processor components.

We present our view of software-based self-testing as a high test quality methodology for low-cost self-testing that achieves its objectives being:

- oriented to *structural testing* adopting well known structural fault models; software-based self-testing has been and can be used with respect to combinational (stuck-at) or sequential (delay) fault models;
- oriented to *component based test generation*, i.e. separate self-test routines are developed for selected processor components, actually for the most important components of the processor first; a significant part of the methodology is spent on prioritization of the processor components according to their importance for testing;

- focused on the *low-cost characteristics* of the self-test routines that are developed, i.e. small size of the self-test programs (code and data), small execution time, small power consumption, all of them under the guidance of the primary goal which is always the highest possible structural fault coverage.

As we will see, component level test generation can be of different flavors depending on the information that is available for the processor core, the individual characteristics of the processor's Instruction Set Architecture, as well as the constraints of a particular application. Therefore, component level test generation can be:

- based on a combinational or sequential ATPG tool which can be guided by a constraints extraction phase executed before ATPG; the extracted constraints describe the effect that the processor's Instruction Set Architecture has on the possible values that can be assigned to processor's components inputs[23];
- based on pseudorandom test patterns sequences which are generated by software-emulated pseudorandom patterns generators[24];
- based on known, pre-computed deterministic test sets for the components of the processor; such pre-computed test sets are available for a set of functional processor components such as arithmetic and storage components.

Moreover, these different approaches for component level self-test generation can be combined together for a specific processor. A subset of the processor's components may be targeted one way and others with an other.

We outline software-based self-test development as a process consisting of four consecutive phases: Phases A, B, C and D, which are analyzed in the following paragraphs, along with visual representation of each of them.

The starting point of software-based self-testing as a low-cost testing methodology is the availability of two pieces of information as a basis for the methodology:

- the Register transfer Level (RTL) description of the processor;
- the Instruction Set Architecture (ISA) of the processor.

ISA is in all cases available in the programmer's manual of the processor and describes the instruction set of the processor and assembly language, the

23 ISA-imposed constraints extraction is not always possible. Obviously, ATPG-based test generation for processor components can be used when a gate-level model of the processor is available.

24 ISA-imposed constraints extraction may be necessary in this case too.

visible registers it includes as well as the addressing modes used for operands access. Detailed knowledge and understanding of the instruction set and assembly language of the processor, consists a key in the successful application of software-based self-testing to it. Particularly, in terms of low-cost testing, "clever" assembly language coding is very important in both cases when self-test code is manually or automatically generated. Moreover, the particular details and restrictions of an instruction set architecture play a significant role on the applicability of software-based self-testing. For example, the same component-level test set can be more efficiently transformed to a self-test routine in the assembly language of one processor than in the assembly language of another. Different instruction sets may lead to different self-test program sizes and execution times.

RTL description of the processor can be either available in a very high level only indicating the most important processor components and their interconnections, or it may be available in more detailed and accurate form if an RTL model of the processor is provided in some hardware description language like VHDL or Verilog. Such a description is sometimes available either in a *synthesizable* form when the processor is purchased by the designer as a soft core, or at least in a *simulatable* form for high level simulation of the processor model when integrated in the SoC architecture. Any RTL description of the processor is useful since it allows an easy way to identify the processor's building components, the existence of which would otherwise be only speculated.

A low-cost software-based self-testing methodology can consist of the following four phases A to D as summarized in Figure 5-7 and further detailed subsequently.

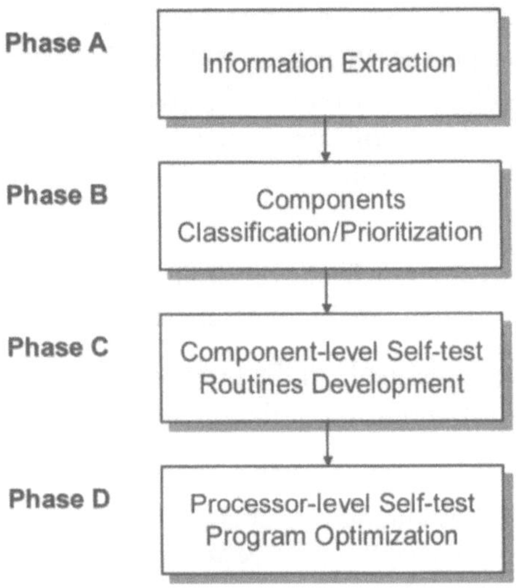

Figure 5-7: Software-based self-testing: overview of the four phases

In the following paragraphs, each of the four phases A through D is presented in a visual way, to enhance readability.

<u>Phase A (Figure 5-8).</u>

During this phase, the RTL information of the processor as well as its Instruction Set Architecture information is used to identify information that will subsequently be used for the actual development of the self-test routines. In particular, during Phase A, the methodology:

- identifies all the components that the processor consists of; this is an essential part of the component-based software-based self-testing because test development is performed at the component level;
- identifies the operations that each of the components performs; and
- identifies instruction sequences consisting of one or more instructions for controlling the component operations, for applying data operands to the operations and for observing the results of operations to processor outputs.

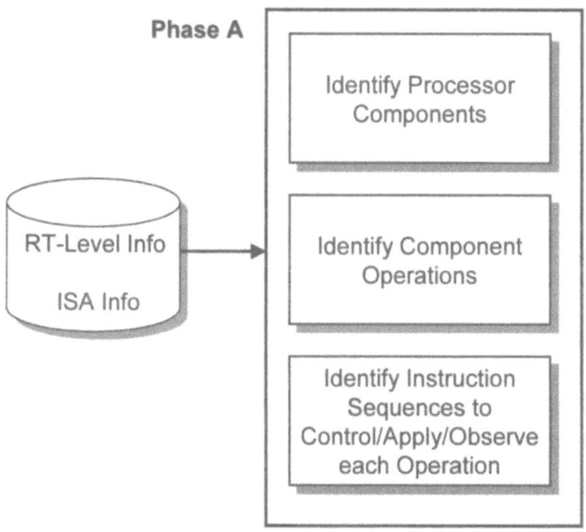

Figure 5-8: Phase A of software-based self-testing.

The final outcomes of Phase A are the following:

- a set of processor components C;
- a set of component operations O_C for each of the components;
- a set of instruction sequences $I_{C,O}$ for the controlling, the application and the observation of operation O to a component C.

These outcomes of Phase A are subsequently used for Phases B, C, and D for the generation of efficient self-test routines for the processor's components.

Phase B (Figure 5-9).

During this second phase of software-based self-testing, processor components (identified in Phase A) are classified in different categories depending on their role in the processor's operation. Each component class/category has different characteristics and different importance in the processor's testability. After classification, the processor components are prioritized to reflect this importance in the overall processor's testability. This prioritization stage is very important in the context of low-cost self-testing because it helps the self-test program development process to quickly reach sufficient fault coverage levels with as small as possible test development effort and with an as small and fast as possible self-test program.

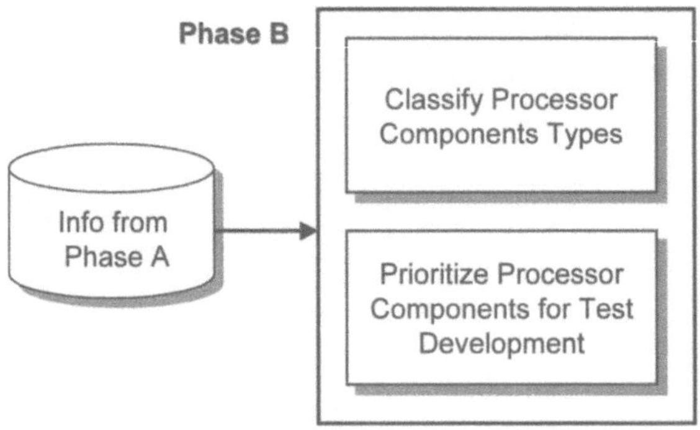

Figure 5-9: Phase B of software-based self-testing.

Phase C (Figure 5-10).

This third phase of software-based self-testing is the actual component level self-test routine development phase. It uses as input the following pieces of information:

- the information extracted during Phase A: set of processor's components, set of operations for each component and instruction sequences for the controlling, application and observation of each component operation;
- the information extracted during Phase B: components classification and components prioritization;
- a *components test library* that contains sufficient sets of test patterns and test generation algorithms in a unified assembly pseudocode for processor components, previously derived using any available method (combinational or sequential ATPG-

generated, pseudorandom-based, known pre-computed deterministic tests, etc)

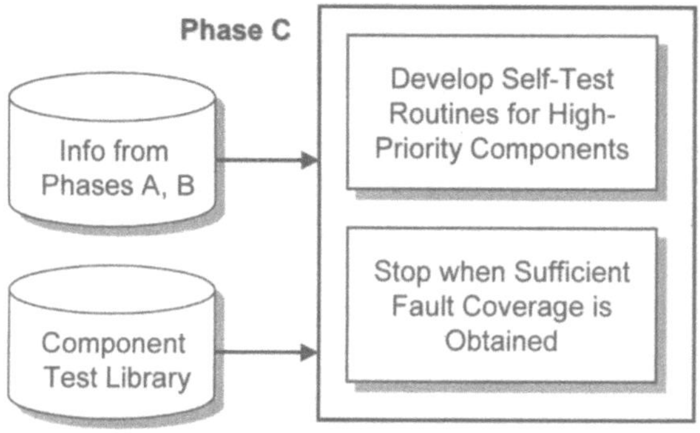

Figure 5-10: Phase C of software-based self-testing.

The objective of Phase C is the development of self-test routines for individual processor's components based on the test sets provided in the component test library. The coding style of the self-test routines may differ from one component to another and also depends on the type of test set or test algorithm available for the component in the component test library (deterministic or pseudorandom).

Component self-test routines development starts from the "most important" components to be tested, i.e. the components that have been previously (during Phase B) identified to have higher priority for self-test routines development. The criteria that determine which components are of higher priority are analyzed in the corresponding subsection below, but basically are the size (gate count) of each of the components and how easily the component is accessible (easy controlling of its inputs and easy observation of its outputs).

Higher priority components are first targeted, because the primary objective of software-based self-testing is to reach high structural fault coverage *as soon as possible*, without even targeting each and every of the processor components.

When test development for one targeted component is completed, the overall processor structural fault coverage is the criterion that determines if software self-test routines development has to stop or to continue. As we will see in the next Chapter where extensive experimental results are given, very high fault coverage can be obtained with just a few important processor components going through self-test routines development. The remaining components are either sufficiently tested as a side-effect or their overall contribution to the missing total fault coverage is very small.

The final result of Phase C of software-based self-testing is a set of component self-test routines developed in the order of their priority and assembled together in a self-test program for the processor. The following code shows just the outline of the overall self-test program for the processor consisting of component's self-test routines.

```
self-test program for processor:
    self-test routine for component 1
    self-test routine for component 2
    ...
    self-test routine for component k
```

After the execution of each self-test routine for a component C_i, $i=1, 2, \ldots, k$, (where $k \leq n$ the number of all processor components) the test responses of all the components have been sent out of the processor in either an unrolled fashion (in several memory words, each one containing a single response to a component test pattern) or in a compacted fashion (one or a few signatures combining all test responses). Test response compaction routines may be either integrated to each of the components' self-test routines, or a global compaction routine may be appended to the processor's self-test program with the task of reading the unrolled test responses from the memory and compacting them in one signature. Details are deiscussed in subsequent section.

Phase D (Figure 5-11).

This last phase of software-based self-testing is an optional one and can be utilized if a particular application has very strict requirements either for smaller size of the self-test program or for smaller self-test execution time. It is the self-test program optimization phase where the self-test routines developed for each of the targeted components in Phase C are optimized according to a set of criteria. The most usual way of self-test routines optimization is the merging of self-test routines developed for different components. Moreover, coding style changes can alter the format of a self-test routine, optimizing it for one criterion.

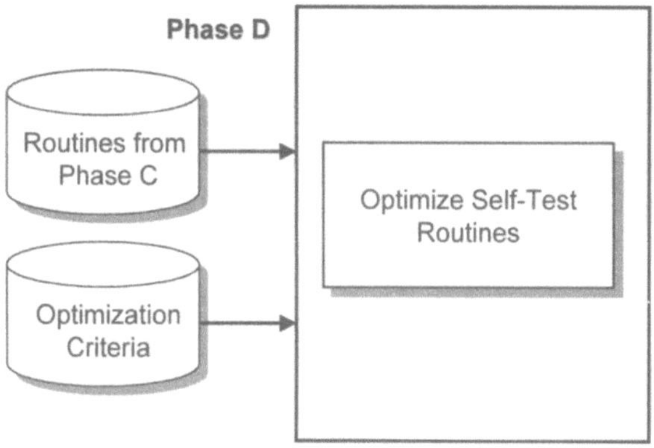

Figure 5-11: Phase D of software-based self-testing.

The following subsections elaborate on the above four different phases of software-based self-testing and in particular on the most critical parts of each of them. The emphasis of the analysis is to justify several choices made in low-cost software-based self-testing. Whenever suitable, examples are given in terms of components types and classes, as well as, in terms of assembly language routines.

The examples throughout this Chapter are simple and informative and are derived using a popular Instruction Set Architecture of a well-known RISC processor, the MIPS RISC processor architecture [22], [85]. Two of the publicly available processor benchmarks that we study in the next Chapter implement the MIPS architecture.

This classical load-store architecture is used to demonstrate several aspects of software-based self-testing, while other different processor architectures are also studied and software-based self-testing is applied to them. Detailed experimental results are presented in the next Chapter.

5.4 Processor components classification

The components that an embedded processor consists of can be classified in different categories according to their use and contribution to the processor operation and instructions execution. In the following subsections, we define generic classes of processor components and elaborate on their characteristics. Figure 5-12 shows the classification of processor components discussed in the following.

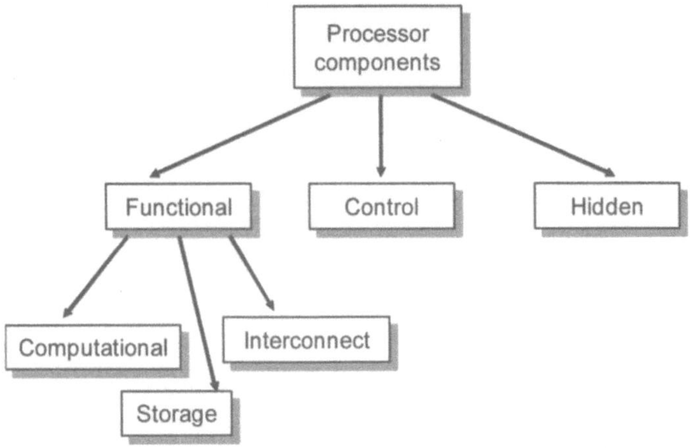

Figure 5-12: Classes of processor components.

5.4.1 Functional components

The *functional components* are the components of a processor that are directly and explicitly related to the execution of processor instructions. The functional components are, in some sense, "visible" to the assembly language programmer, or in other words their existence is easily implied by the instruction set architecture of the processor. Functional components may belong to one of the following sub-classes:

- *Computational functional components* which perform specific arithmetic/logic operations on data as instructed by the functionality of processor instructions. Such components are: Arithmetic Logic Units (ALUs), adders, subtracters, comparators, incrementers, shifters, barrel shifters, multipliers, dividers, or compound components consisting of some of the previous. The computational functional components that realize arithmetic operations may either deal with integer or floating point arithmetic.
- *Storage functional components* which serve as storage elements for data and control information. Storage components contain data that is fed to the inputs of computational functional components or is captured at their output after their computation is completed. This sub-class of storage functional components includes all assembly programmer visible registers, accumulators, the register file (s), several pointers and special processor registers that store control and status information visible to the assembly programmer.

- *Interconnect functional components* which are components that implement the interconnection between other types of processor functional components and control the flow of data in the processor's datapath, mainly between the previous two sub-classes of computational and storage functional components. Interconnect components include multiplexers controlled by appropriate control signals generated after instruction decoding, or bus control elements like tri-state buffers realizing the same task.

The information on the number and types of the processor's functional components can be directly derived from the RTL description of the processor or, in the worse case, can be implied by the programmer's manual and instruction set. Simply stated, the existence of the functional components analyzed above is easily implied by a careful study of the Instruction Set Architecture of the embedded processor. We outline some simple examples to clarify this statement.

If the instruction set of the processor includes an integer multiplication instruction, this means that the processor contains a hardware multiplier of either serial, multi-cycle architecture or of a parallel, single-cycle architecture. The existence of the multiplication instruction itself delivers this information, even if an RTL description of the processor is not available. The integer multiplier, in this case, is a computational functional component. For example, in the MIPS instruction set the integer multiply instruction[25]:

```
mult Rs, Rt
```

implies that an integer multiplier component exists in the processor to realize the multiplication of the contents of the 32-bit general purpose registers `Rs` and `Rt` (both of the integer register file of the processor). According to the MIPS instruction set architecture description, the product of the multiplication is stored in two special 32-bit integer registers named `Hi` and `Lo`, not explicitly mentioned in the syntax of the multiplication instruction. If the instruction takes 32 cycles to be completed – information available in the programmer's manual – this means that the multiplier component implements serial multiplication, while if the instruction takes 1 cycle to be completed, the multiplier is a parallel one[26].

[25] In the examples of this Chapter, we use the assembly language of the MIPS processors. We denote the registers as Rs, Rt, Rd or as R1, R2, etc, for simplicity reasons, although traditionally, these registers are denoted as \$s0, \$s1, etc in the MIPS assembly language.

[26] Multiplication using a parallel multiplier may take more than 1 clock cycle, if for performance reasons of the other instructions, the multiplication is broken in more than one phases and takes more (usually two) clock cycles.

As a second example we consider the case in which the instruction set contains an arithmetic or logic operation where the second, for example, operand of the operation (in the assembly language description) can be a general purpose register (from the register file), a memory location or an immediate operand (coming directly from the instruction itself). In this case, a multi-input (three-input in this case) multiplexer should exist at the second operand input of the component that performs the operation. This multiplexer is an interconnect functional component which existence is easily extracted from the instruction set architecture even if an RTL description is not available. For example, in the MIPS instruction set architecture, the following two instructions perform the bitwise-OR operation between two general purpose registers (first instruction) or a register and an immediate operand (second instruction). In both cases, the result is stored in a general purpose register `Rd`. The second operand of the operation is in the first case another register `Rt` and in the second case it is an immediate operand directly coming from the instruction itself (the immediate operand `Imm` in the MIPS machine language format consists of 16 bits).

```
or  Rd, Rs, Rt
ori Rd, Rs, Imm
```

A two-input multiplexer that feeds the second input of the ALU which implements the bitwise-OR operation is in this case an identified interconnect functional component.

Storage functional components are the easiest case of the three sub-classes of functional components since in most cases their existence is directly included in the assembly language instructions or explicitly stated in the programmer's manual (like in the case of the `Hi` and `Lo` registers of MIPS which although are not referred in assembly language are explicitly mentioned in the programmer's manual of the processor). Usually, an accumulator's or general purpose register's name is part of assembly language format, while also a status register that saves the status information after instructions are executed, is also directly implied by the format and meaning of special processor instructions used for its manipulation. Getting, again, an example from the MIPS instruction set architecture, the following assembly language instruction:

```
and R4, R2, R3
```

identifies three general purpose registers from the integer register file: `R2`, `R3` and `R4`. Moreover, an additional piece of information that can be extracted from the existence of this instruction is that the processor contains a register file with at least 2 read ports and at least 1 write port.

In data-intensive processor architectures where data processing parallelism is critical, like in the case of Digital Signal Processors (DSPs), the class of functional processor components dominates the processor circuit size more than in any other case of general purpose processor or controller. This is true for example, because more than one computational functional components of the same type may co-exist to increase parallelism, or because more general purpose registers than in typical processors exist to increase the available storage elements in the DSP architecture.

Functional components and in particular the computational sub-class of functional components are usually large components in size and thus in number of faults. In modern bus widths of 32 and 64 bits, and in DSP applications with increased precision requirements and larger internal busses, the functional units that perform calculations like addition, multiplication, division or combined calculations like multiply-accumulate in DSPs, are very large in size and consist of some tens of thousands of logic gates. Such components have high importance in software-based self-test routines development as we will see in a while after completing the description of the different classes of processor components.

5.4.2 Control components

The *control components* are those that control either the flow of instructions and data inside the processor or the flow of data from and to the external environment (memory subsystem, peripherals, etc).

A classical control component is the component that implements the instruction decoding and produces the control signals for the functional components of the processor: the processor control unit. The processor control unit may be implemented in different ways: as a Finite State Machine (FSM), or a microprogrammed control unit. If the processor is pipelined a more complex pipeline control unit is implemented.

Other typical control components are the instruction and data memory controllers that are related to the task of instruction fetching from the instruction memory and are also related to the different addressing modes of the instruction set.

The common characteristic of control components is that they are not directly related to specific functions of the processor or directly implied by the processor's instruction set. Control components existence is not evident in the instruction format and micro-operations of the processor and the actual implementation of the control units of a processor may significantly differ from one implementation to another. On the contrary, the functional components of different processors are implemented more or less with the same architecture.

The control components of a processor are usually much smaller in size (gate count) compared to the functional components, but their testing is also important because if they malfunction it is very unlikely that any instruction of the processor can be correctly executed. Control components are moreover, more difficult to be tested compared with functional components because of their reduced accessibility and variety of internal implementations.

With the scaling of processor word lengths towards 32 and 64 bits functional components have grown in size while control components have a smaller increase.

5.4.3 Hidden components

The *hidden components* of the processor are those that are included in a processor's architecture usually to increase its performance and instruction throughput.

The hidden components are not visible to the assembly language programmer and user programs should functionally operate the same way under the existence or absence of the hidden components. The only difference in program execution should be its performance since in the case of existence of hidden components performance must be higher than without them.

A classical group of hidden components is the one consisting of the components that implement pipelining, the most valuable performance-enhancing technique devised the last decades to increase processor's performance and instruction execution throughput. The hidden components related to pipelining include the pipeline registers between the different pipeline stages[27], the pipeline multiplexers and the control logic that determines the operation of the processor's pipeline. These include components involved with pipeline hazards detection, pipeline interlock and forwarding/bypassing logic.

The hidden components of a processor may be either of storage, interconnect or control nature. Storage hidden components have a similar operation with the sub-class of storage functional components. Pipeline registers belong to this type of storage hidden components. The pipeline control logic has control characteristics similar to the control components class. Finally, there are hidden components that implement pipelining which have an interconnect nature. These are the multiplexers of the pipeline structure of a processor which realize the forwarding of data in the pipeline

[27] Pipeline registers do not belong to the class of storage functional components because they are not visible to the assembly language programmer.

when pipeline hazards are detected. The logic which detects the existence of pipeline hazards consists of another type of hidden components, the pipeline comparators which can be considered part of the pipeline control logic.

Other cases of hidden components are those related to other performance increasing mechanisms like Instruction Level Parallelism (ILP) and speculative mechanisms to improve processor performance such as branch prediction schemes. Such prediction mechanisms can be added to a processor to improve programs execution speeds but their malfunctioning will only lead to reduced performance and not to functional errors in programs execution.

It is obvious from the description above that self-test routines development for hidden processor's components (or any other testing means) may be the most difficult among the different components classes. The situation is simplified when the processor under test does not contain such sophisticated mechanisms (processor without pipelining). This is true in many cases today where previous generations microprocessors are being implemented as embedded processors. In a large set of SoC designs, the performance delivered by such non-pipelined embedded processors is sufficient, and therefore software-based self-testing for them has to deal only with functional and control components.

In cases that the performance of classical embedded processors is not enough for an application, modern embedded processors are employed. The majority of modern embedded processors include performance-enhancing mechanisms like a *k*-stage pipeline structure (*k* may range from 2 to 6 or even more in embedded processors).

Although direct self-test routines development for hidden components is not easy, the actual testability of some of them (like the pipeline-related components) when self-test routines are developed for the functional components of the processor can be inherently very high. Intuitively, this is true due to the fact that pipeline structure is a "transparent" mechanism that is not used to block the execution of instructions but rather to accelerate it.

The important factor that will determine how much of test development effort and cost will be spent on pipelining, is their relative size and contribution to overall processor's testing.

5.5 Processor components test prioritization

It has been pointed out so far, that software-based self-testing is considered as a low-cost test methodology for embedded processors and corresponding SoC designs that contain them. Therefore, the primary goal of reaching high structural fault coverage must be achieved at the expense of an as low as possible engineering effort and cost for test generation and as small as possible test application time. In software-based self-testing, the low-cost test target can be achieved when *small* and *fast* test programs are developed in the *smallest* possible test development cost. Towards this aim, the processor components previously classified in the three classes of functional, control and hidden must be prioritized in terms of their importance for test generation. Self-test program development for each component will start from the most important components and will then continue to other components with a descending order of importance. Component level self-test routines development continues until sufficient fault coverage is reached. The following flow diagram of Figure 5-13 depicts this iterative process of software-based self-testing.

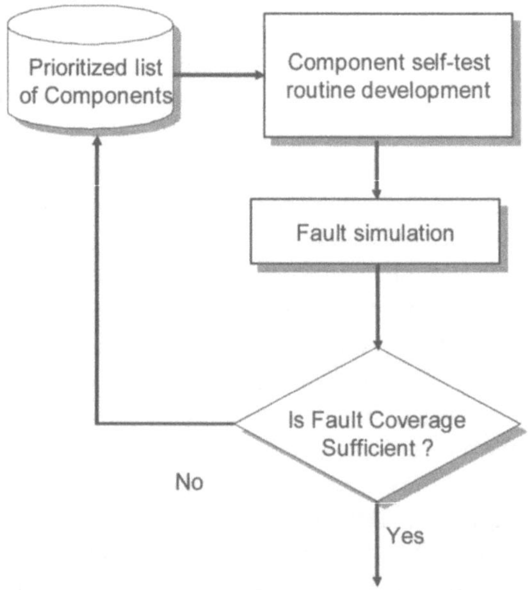

Figure 5-13: Prioritized component-level self-test program generation.

The prioritization criteria for a processor's components are analyzed in this subsection. After prioritization is finished, component level self-test routines development is performed. The components with higher priority are first targeted and self-test routines are developed for them. Fault simulation evaluates the overall obtained fault coverage and determines if further test development is necessary to reach the required fault coverage for the processor. A gate level netlist of the processor is required for gate level fault simulation. We must note that a gate level netlist at this point is necessary only to decide if test development must continue. The component level test development process on the other side may or may not need such a piece of information to be available as we see later.

The criteria that are used for component prioritization for low-cost software-based self-testing are discussed and analyzed in the subsequent paragraphs. The criteria are in summary the following:

- Criterion 1 – component *size* and *percentage* of total processor faults set that belongs to the component.
- Criterion 2 – component *accessibility* and *ease* of testing using processor instructions.
- Criterion 3 – *correlation* of the component's testability with the testability of other components.

We elaborate on the importance of the three criteria in the following subsections.

5.5.1 Component size and contribution to fault coverage

This criterion, simply stated, gives the following advice: component level self-test routines development should give higher priority to *large* components that contain large number of faults.

When a processor component is identified from the instruction set architecture and the RTL model of the processor, its relative size and its number of faults as a percentage of the total number of processor faults is a valuable piece of information that will guide the development of self-test routines. Large processor components containing large numbers of faults must be assigned higher priority compared to smaller components because high coverage of the faults of this component will have a significant contribution to the overall processor's fault coverage. For example, developing a self-test routine that obtains a 90% fault coverage for a processor component that occupies the 30% of the processor gate count (and faults number), contributes a 30% × 90% = 27% to the total processor fault coverage, while a 99% fault coverage on a smaller component that occupies only 10% of the processor gate count (and faults as well) will only contribute by 10% × 99% = 9.9% to the overall processor fault coverage. Needless to note, that reaching 90% fault coverage in a component is always much easier than to reach 99% in another.

This criterion, although simple and intuitive, is the first one to be considered if low-cost testing is the primary objective. Large components must definitely be given higher priority than smaller ones since their effective testing will sooner lead to large total fault coverage.

The actual application of this criterion for the prioritization of the processor's components requires that the information for the gate count of the processor components is known. Unfortunately, this information is not always available to the test engineer[28]. In this case, software-based self-test development can be only based on speculations for the relative sizes of the processor components.

Two speculations are in almost all cases true and can be easily followed as a guideline for test development:

- functional components of all sub-classes (computational, interconnect and storage) are larger in size and thus is faults count than control and hidden components; therefore, they should be assigned higher priority for self-test routines development;
- among the three types of functional components those with the largest size are the computational ones that perform arithmetic operations like adders, ALUs, multipliers, dividers and shifters and also the storage components, the most important of which are the register files.

A safe way to verify the correctness of the above statements is by providing data from representative processors and their components gate counts. The experimental results presented in Chapter 6 are towards this aim. As a first indication we mention that, depending on the synthesis library and the internal architecture of the components, the computational functional components of the Plasma/MIPS processor model [128] occupy from 24.06% up to 48.98% of the total gate count of the processor. The computational functional components in this processor model include an Arithmetic Logic Unit (ALU), a multiplier (that can be either a serial or a parallel one), a divider (in serial implementation) and a shifter component. Moreover, the register file of the embedded processor occupies from 37.98% up to 56.74% of its total gate count. Combined, the computational functional components and the register file (a storage functional component) of the processor occupy at least 80.89% and as much as 87.72% of the total processor area. Again, the different percentages depend on the synthesis library, the synthesis options and also the different internal architectures of the processor components (e.g. parallel vs. serial multiplier). The total number of gates in the several different implementations of Plasma range between 17,500 and 31,000 gates.

The above gate counts and corresponding percentages give a clear indication of the importance of the computational functional components and

[28] Gate counts are available either when a gate-level netlist of the component is given or when a synthesizable model of the processor (and thus the components) is available for synthesis.

the storage functional components and at least the first reason for which they should be first addressed by software-based self-test routines development.

5.5.2 Component accessibility and ease of test

The second, but equally important with the first one, criterion for prioritization of processor components for low-cost self-test program generation, is the component's accessibility from outside the processor using processor instructions. The higher this accessibility is, the easier the testing of the component is.

The development of self-test routines is much easier and requires less engineering effort and cost when the component under test is easily accessible by programmer-visible, general purpose registers of the processor. This means that the component inputs are connected to such registers and component outputs drive such registers as well.

In this case, the application of a single test pattern to the component under test by a self-test program simply consists of the following three steps:

- execute instruction(s) to load input register(s) with test pattern from outside (data memory or instruction memory[29]);
- execute an instruction to apply the test pattern to the component; and
- execute instruction(s) to store the result from register(s) to outside the processor (memory)

These three steps correspond to the software-based self-testing steps we first mentioned in Figure 5-3. A simple example below shows such an easy application of a test pattern to a shifter component of a MIPS-like embedded processor.

```
lw R2, offset1(R4)
lw R3, offset2(R4)
srlv R1, R2, R3
sw R1, offset3(R4)
```

The component under test is a shifter and the operation tested by this portion of assembly code is the right logical shift of the shifter. Register R2 contains the original value to be right-shifted (first part of the test pattern to be applied to the shifter) and register R3 contains the number of positions for right-shifting (second part of the test pattern for the shifter). In our example, both these values are derived from memory (base address contained in

[29] When the immediate addressing mode is used, test patterns are actually stored in the instruction memory, i.e. as part of the instructions.

register R4 incremented by the amount offset1 or offset2, respectively). The loading of these two values is done by the first two lw (load word) instructions. The third instruction of the code, applies the test pattern to the shifter (srlv = shift right logical by variable; the shift amount is stored in a variable/register) and the shifted result (the output of the shifter) is stored into register R1. Finally, the last instruction (sw – store word), stores the content of register R1 (the component's test response) to a memory location outside the processor. In the sw instruction the memory location where the test response will be stored is again identified by a base register content (R4) and an offset to be added to it (offset3).

Such assembly language code portions for the application of component level test patterns can be constructed only for components which have direct access from/to general purpose registers. In most cases such components are only the *computational functional components* of the processor such as the shifter component discussed in the example. The inputs of the computational functional components are directly accessible by programmer visible register(s) which in turn can be easily loaded with the desired values (test patterns) and also the components outputs (result of the calculation – test response) are driven to other programmer visible register(s) which values in turn can be easily transferred outside of the processor and stored to memory locations for further evaluation.

Equally well accessible and easily testable are most of the storage functional components of the processor because of the availability of processor instructions that directly write values to such components (like general purpose registers, accumulators, etc) and also instructions that directly read their values and transfer them out of the processor. For example, a test pattern can be applied to the general purpose register R6 of a MIPS-like processor with the following lines of assembly language code[30].

```
li R6, test-pattern
sw R6, offset(R1)
```

[30] We remind that Ri is not the usual notation of registers in MIPS assembly language, but rather, it is $s0, $s1, $t0, $t1, etc. We use the Ri notation for simplicity.

The first instruction is the load immediate (`li`) instruction[31] which loads the register `R6` with the test pattern value to be applied to the component under test: register `R6`. The `li` instruction does not apply a test pattern which is stored in data memory (as an `lw` instruction does) but a test pattern which is stored in instruction memory (the instruction itself). The second instruction (`sw`) stores the content of the register `R6` (which is now the component's test response) to a memory location addressed by the base address in register `R1` increased by the `offset`.

The same simplicity in self-test program development and application does not apply to other components than computational and storage functional components because (a) they are either not connected to programmer-visible registers but are rather connected to special registers not visible to the programmer; and/or (b) they can not be directly accessible by processor instructions for test pattern application.

Therefore, computational and storage functional components must be given priority for self-test routine development since they can quickly contribute to a very high fault coverage for the entire processor with simple self-test routines. If we combine this second criterion of component accessibility and ease of testing with the previous criterion of relative component size, we can see that functional components are very important for self-test program development in a low-cost software-based self-testing approach. The third criterion discussed in the following subsection supports further this importance.

5.5.3 Components' testability correlation

The third criterion is related to the ability to test some components as a side effect of testing other ones. This means that when a set of test patterns is applied to a processor component to obtain sufficient fault coverage, another processor component is tested to some degree as well. Let us examine situations where this applies.

- When a functional component (a computational or a storage one) is being tested by an instruction sequence specially developed for

[31] Actually, the `li` instruction is not a real machine instruction of the MIPS architecture but rather an assembler pseudo-instruction (also called macro) that the aseembler decomposes to two instructions `lui` (load upper immediate) and `ori` (or immediate). Load upper immediate loads a 16-bit quantity to the high order (most significant) half of the register and `ori` loads the low order 16 bits of it. Therefore, the instruction `li R6, test-pattern` is translated to:

```
lui R6, high-half
ori R6, R6, low-half
```

where `test-pattern=high-half & low-half` (`&` denotes concatenation).

this purpose, then part of the control logic of the processor is also tested in parallel. Such part is for example the instruction decode part of the control logic. In particular, when the instruction sequence that tests the functional component contains a sufficient variety of processor instructions then it is likely that significant fault coverage for the instruction decode part is obtained as well.

- When a functional component is being tested, then part of interconnect functional components is also tested in parallel. Multiplexers at the inputs or the outputs of the functional component under test will be partially tested.
- When a functional component is being tested, then part of the pipeline logic of the processor is also tested in parallel. At least parts of the pipeline registers and parts of the multiplexers of forwarding paths will be tested to some extent. Therefore, if the instruction sequence that tests the functional component contains instructions that pass through different paths of the pipeline forwarding logic, then the pipeline logic is also sufficiently tested.

We remark at this point that the criteria described in this and the previous subsections are only meant to be used to prioritize the importance of processor components for test development and are not in any sense statements that are absolutely true in any processor architecture and implementation. These criteria are good indications that some components must be given priority over others for a low-cost development of self-test routines.

Concluding with this third criterion, we mention that when other components than functional components are being tested, for example control components, it is very unlikely that other components are being sufficiently tested as well. For example, when executing a self-test program for the processor instruction decode component, what is necessary is to pass through all (or most) of the different instructions of the processor. When such a self-test program is executed only a few of its instructions detect also some faults in a computational functional unit, like an adder, which requires a variety of data values to be applied to it (the adder). Therefore, sufficient testing of the decoder component does not give, as a side effect, sufficient fault coverage for the adder component or other functional components. On the contrary, when sufficient fault coverage is obtained by a self-test routine for the adder component then, in parallel, the decode unit is also sufficiently tested at its part that is dedicated for the decoding of the addition-related instructions.

In the global view, when separate self-test routines have been developed for all the processors functional units (or at least for the computational and the storage components) then the other components (control components like the instruction decode and the instruction sequencing components and hidden components like the pipeline registers, multiplexers and control logic) are also very likely to be sufficiently tested. The opposite situation is not true: when self-test routines have been developed targeting the control components or the hidden components then the functional components are not sufficiently tested as well, simply because the variety of data required to test them is not included in the self-test routines for the control and hidden components.

After having described the criteria for the prioritization of the processor components for self-test routines development, we elaborate in the next two subsections on the identification and selection of component operations to be tested, as well as, on the selection of appropriate operands to test the selected operations with processor instructions.

5.6 Component operations identification and selection

Phase A of software-based self-test development identifies a set of instruction sequences $I_{C,O}$ which consists of processor instructions I that, during execution cause component C to perform operation O. The instructions that belong to the same set $I_{C,O}$ have different controllability/observability properties since, when operation O is performed, the inputs of component C are driven by internal processor registers with different controllability characteristics while the outputs of component C are forwarded to internal processor registers with different observability characteristics.

Different controllability and observability for processor registers refers to the ease of writing values to a register and transferring its contents to the outside of the processor. Higher controllability means that a smaller number of processor instructions are required to assign a value to a register; while higher observability means that a smaller number of instructions are required to transfer the contents of the register out of the processor.

Therefore, for every operation O_C of a component C, derived in phase A, an appropriate instruction sequence I must be selected from the set of instruction sequences $I_{C,O}$ that apply the same operation to the same component. The objective is to end up with the shortest instruction sequence required to apply any particular operand to the component inputs and the shortest instruction sequence required to propagate the component outputs to the processor primary outputs.

Let us see an example to demonstrate the instruction selection for the computational functional component Arithmetic Logic Unit (ALU) of the MIPS-like architecture. Such an ALU has two 32-bit inputs, one 32-bit output and a control input which specifies the operations that the component performs under the control of the processor instructions. Figure 5-14 shows the ALU component and its inputs and outputs. The operands are 32-bits wide, the result is also 32-bits wide and the operation (control) input consists of 3 bits (there are eight different operations that the component performs).

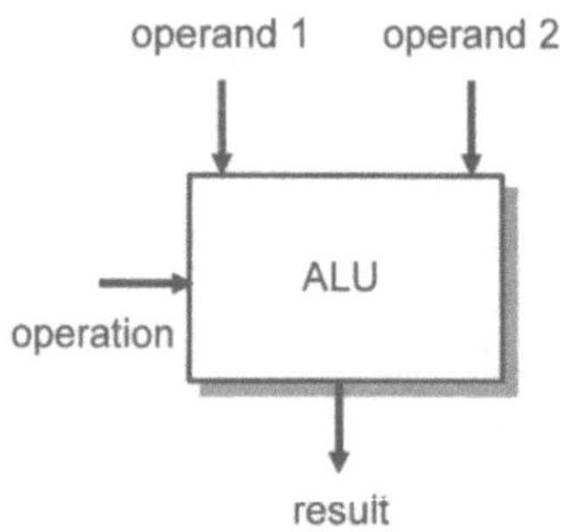

Figure 5-14: ALU component of the MIPS-like processor.

The set of operations O_{ALU} that the ALU component performs is the following:

```
OALU = { add,
         subtract,
         and,
         or,
         nor,
         xor,
         set_on_less_than_unsigned,
         set_on_less_than_signed
}
```

For every operation O belonging to the O_{ALU} set, we identify the corresponding set of processor instructions $I_{ALU,O}$ that during execution cause the ALU component to perform operation O. As the ALU can perform 8 operations, the sets of instructions are the following:

$I_{ALU,ADD}$

$I_{ALU,SUBTRACT}$

$I_{ALU,AND}$

$I_{ALU,OR}$

$I_{ALU,NOR}$

$I_{ALU,XOR}$

$I_{ALU,SET\_ON\_LESS\_THAN\_UNSIGNED}$

$I_{ALU,SET\_ON\_LESS\_THAN\_SIGNED}$

For the development of self-test routine for the ALU, one instruction I from each set $I_{ALU,O}$ (eight different sets for the eight different operations of the ALU) is needed to apply a data operand for each operation O.

For example, the set $I_{ALU,NOR}$ has only one processor instruction and thus the selection is straightforward. The following instruction can only be used to test for the NOR operation of the ALU:

```
nor Rd, Rt, Rs
```

The sets of instructions $I_{ALU,OR}$, $I_{ALU,XOR}$, $I_{ALU,AND}$, $I_{ALU,SUBTRACT}$, $I_{ALU,SET\_ON\_LESS\_THAN\_UNSIGNED}$, $I_{ALU,SET\_ON\_LESS\_THAN\_SIGNED}$, all consist of two instructions, one in the R-type (register) format of MIPS where the operation is applied between two registers of the processor and the other in the I-type (immediate) format of MIPS where the operation is applied between a register and an immediate operand.

The instructions in the I-type format have less controllability than the instructions in the R-type format. Thus, the instructions in the R-type format must be selected because they provide the best controllability and observability characteristics due to the use of the fully controllable and observable general purpose registers of the register file. Therefore, from these sets, the following instructions will be selected to test the corresponding operation of the ALU.

```
or Rd, Rt, Rs
xor Rd, Rt, Rs
and Rd, Rt, Rs
sub Rd, Rt, Rs
subu Rd, Rt, Rs
slt Rd, Rt, Rs
sltu Rd, Rt, Rs
```

Finally, the $I_{ALU,ADD}$ set of instructions consists of a large number of instructions since the ALU is also used in memory reference instructions that calculate sums of a base register content and an offset. Instructions included in this set are the following among others.

```
add Rd, Rs, Rt
addu Rd, Rs, Rt
addi Rt, Rs, Imm
addiu Rt, Rs, Imm
lw Rt, offset(Rs)
sw Rt, offset(Rs)
```

In this case, the selected instruction would be any of the first two of the listed instructions above that belongs to the R-type format because it possesses the best controllability and observability since it uses the general purpose registers of the register file of the processor.

5.7 Operand selection

As we saw in the previous subsection, each of the processor components performs a set of different operations under the control of the processor instructions. Of course, arithmetic and logic computational functional components have the largest variety of operations, like in the case of a multi-functional ALU. The implementation of the different operations can be done internally in many different ways, such as separate implementation of the arithmetic and logic operations and their combination with a set of multiplexers. At the instruction set level, and under the assumption that the only available structural information of the processor may be the RT level description of it, software-based self-testing concentrates on the application of a set of test patterns at the component operand inputs so that any particular operation is sufficiently tested.

To outline an example, we can again consider the case of the MIPS ALU with the operations described in the previous subsection. Each of these operations excites a different part of the processor ALU as Table 5-1 shows.

Operation	ALU part used
`add`	Arithmetic part (adder)
`subtract`	Arithmetic part (subtracter)
`and`	Logic part (and)
`or`	Logic part (or)
`nor`	Logic part (nor)
`xor`	Logic part (xor)
`set on less than unsigned`	Arithmetic part (subtracter)
`set on less that signed`	Arithmetic part (subtracter)

Table 5-1: Operations of the MIPS ALU.

We see in Table 5-1 that each operation excites a different part of the ALU component (in this case that the component is an ALU, some operations excite the arithmetic and some others excite the logic part of it). Different sets of test patterns are required to excite and detect the faults in these different parts of the component.

Appropriate selection of component-level test patterns is an essential factor for the successful development of self-test routines for components. In this subsection the focus is on this aspect of software-based self-testing: operand selection.

Component operand selection corresponds to component-level test pattern generation and therefore the different styles of test pattern generation will be detailed. These different styles are discussed in the subsequent paragraphs and will be related right after with the corresponding coding styles for software-based self-testing.

5.7.1 Self-test routine development: ATPG

According to this approach, test generation for the processor components is based on the use of a combinational or sequential Automatic Test Pattern Generator (ATPG), depending on the type of component considered.

Application of ATPG-based self-test routines development requires the availability of a gate level model of the processor (or at least for the component under test) on which the ATPG will operate. Alternatively, a synthesizable model of the processor may be used for gate-level ATPG after synthesis has been performed on it. In the case, that neither a gate-level netlist nor a synthesizable model of the processor is available (the processor has been delivered as a hard core) ATPG-based self-test routine development can't be supported.

ATPG-based test development for processor components may be or may not be successful, depending on the complexity and size of the component under test. In the case of simple combinational components, combinational ATPG is usually successfully applied. On the contrary, in components with a considerable sequential depth and with a large number of storage elements, sequential ATPG algorithms and tools may be unable to reach sufficient fault coverage for the component.

Moreover, as studied in many approaches presented in the literature, component-level ATPG may require *constraint extraction* so that the derived component test patterns can be applied to them using processor instructions. Constraints that must be extracted are *spatial* constraints and/or *temporal* constraints.

Constraint extraction is a promising direction related to processor testing and in general sequential ATPG for large hierarchical designs, like processors, but it is still under development and it is very likely that it will lead to significant results and tools the next few years. Conceptually, constraints extraction is a *bottom-up* approach, since low level constraints are extracted for processor components and then transferred upwards to high level processor instructions (called also *realizable* tests). This bottom-up approach, although theoretically correct, does not always obtain sufficient results. Constraints are either very difficult to be extracted, but even if this is the case, mapping of the constraints to processor instructions is not a straightforward task.

In the case that a gate level model of the component is available (or it can be obtained from synthesis) and combinational or sequential ATPG succeeds to produce a test set of sufficient fault coverage (with or without the assistance of constraints extraction), the result is a set of *k* component test patterns:

```
atpg-test-pattern-1
atpg-test-pattern-2
...
atpg-test-pattern-k
```

The two important properties of the set of *k* test patterns for self-test development are:

- the cardinality of the test set (number *k*), and
- their format and correlation.

A self-test routine that tests a processor component `C` for one of its operations `O`, using an instruction `I` selected among the instructions of the set $I_{C,O}$ and applying a test set of *k* test patterns is outlined below. We assume, as in all places in this Chapter, a classical RISC load-store architecture where arithmetic and logic operations are applied only to general purpose registers.

```
atpg-tests-loop:
    load register(s) with pattern(s) from memory
    apply instruction I
    store result(s) to memory
    repeat atpg-tests-loop
```

The software loop in the above pseudocode will be repeated *k* times where *k* is the total number of test patterns generated by the ATPG. In this self-test routine style, the test patterns of the component have been stored in *data memory* (as variables in the assembly language source code) and a loop applies them consecutively to the component under test.

Alternatively, the component test patterns can be applied to it with another style of self-test code. In this second style, the test patterns are stored in the instruction memory of the processor (the instructions themselves) and are applied using the *immediate* instruction format of the processor (also called immediate addressing mode). The following pseudocode outlines this self-test style for the application of ATPG-based test patterns.

```
atpg-tests-no-loop:
      load register(s) with immediate pattern 1
      apply instruction I
      store result(s) to memory
      load register(s) with immediate pattern 2
      apply instruction I
      store result(s) to memory
      ...
      load register(s) with immediate pattern k
      apply instruction I
      store result(s) to memory
```

In the loop-based application of test patterns, they occupy part of the data segment (thus data memory) of the self-test program as variables. In the second case, the test patterns are not stored in variables (data memory) but they rather occupy part of the code segment (instruction memory) of the self-test program.

As an example, consider that the functional component under test is a binary subtracter, a gate-level netlist of the component is available and also that the ATPG has generated a test set consisting of k test patterns. When the ATPG-based test patterns are applied in a loop from data memory then the following code of MIPS assembly language gives an idea of how the k test patterns can be applied to the subtracter[32].

```
test-subtracter: andi R4, R4, 0
next-test:       lw R2, xstart(R4)
                 lw R3, ystart(R4)
                 sub R1, R2, R3
                 sw R1, rstart(R4)
                 addi R4, R4, 4
                 slti R5, R4, 4*k
                 bne R5, R0, next-test
```

Since the subtracter is a two-operand computational functional component, the k test patterns that the ATPG generates have an X operand part and a Y operand part. We assume that the k patterns are stored in data memory in consecutive memory locations starting at `xstart` and `ystart` addresses for the X operand and the Y operand, respectively. Register `R4`

[32] `andi`, `addi` and `slti` are the *immediate* versions of the logic and, addition and set-on-less-than instructions. Set-on-less-than sets the first register to 1 if the second operand is less than the third. Branq if not equal (`bne`) instruction takes the branch if the compared registers are not equal. Finally register `R0` (denoted `$zero` in MIPS assembly) has always an all-zero content.

counts the number of test patterns times four[33] to show the correct addresses of data memory for loading the test patterns applied to the subtracter and storing the test response. Register `R4` also controls the number of repetitions of the loop (k repetitions in total). Registers `R2` and `R3` are loaded with the next pattern in each repetition and the result of the subtraction is put by the subtracter in the `R1` register. The `R1` register that contains the test response of the subtraction is finally stored to an array of consecutive memory locations starting at address `rstart` as shown in the code. At the end of each loop iteration, the counter `R4` is incremented by 1 and a check is performed to see if the k test patterns have been exhausted. If this is not the case, the loop is repeated.

The self-test code above consists of eight instructions (words) and $2k$ words storing the k two-word test patterns (one word for operand X and one word for operand Y). The execution time of this routine depends on the number k of test patterns to be applied to the subtracter. The exact execution time and number of clock cycles depends on whether the processor implementation is pipelined or not and also on the latency of memory read and write cycles.

For simplicity, le us consider a non-pipelined processor. If we assume that each instruction executes in one clock cycle apart from memory reads and writes which take 2 clock cycles, then a rough estimation of the number of clock cycles required for the completion of above self-test routine is $10k$ clock cycles[34].

Figure 5-15 presents in a visual way the application of ATPG-generated test patterns from a data memory array using load word instructions. Two instructions are required to load the registers with the test patterns. The test vectors are then applied by the subtract instruction to the subtracter component as the code above shows.

[33] In the case of MIPS, we consider words of 4 bytes and hence addresses of sequential words differ by 4.

[34] In a pipelined implementation, one instruction will be completed in each cycle but with a smaller period. Pipeline and memory stalls will increase the execution time of each loop.

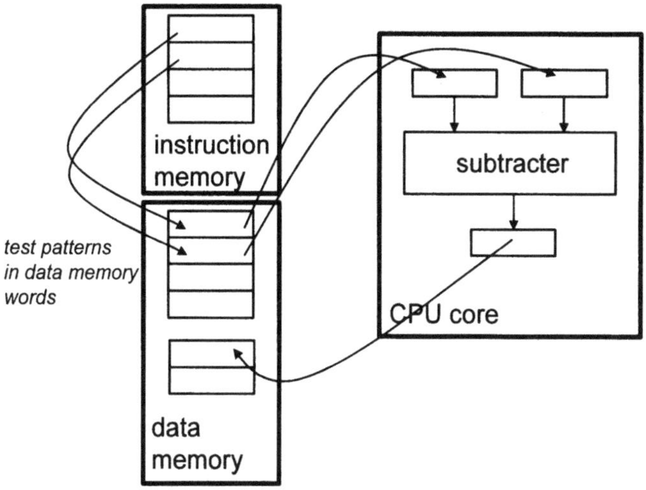

Figure 5-15: ATPG test patterns application from memory.

The alternative way to implement the self-test code for the application of the k test patterns to the subtracter is to use the immediate operand addressing mode that all (or most) processors' instruction set architectures include. According to the immediate addressing mode, an operand is part of the instruction. The MIPS architecture consists of 32-bit instructions where the immediate operand can occupy a 16-bit part of the instruction. For example, in the immediate addressing mode instruction:

```
andi Rt, Rs, Imm
```

which implements the logical AND operation between register `Rs` and immediate operand `Imm` and stores the result in register `Rt`, the immediate operand `Imm` is a 16-bits long.

In order to store a 32-bit value to a register using 16-bit immediate operands an additional instruction is used in MIPS which loads the upper half 16-bit part of the register. This instruction is called *load upper immediate* (`lui`) and loads the 16-bits immediate value to the upper 16 bits of the register `Rt` (bits 31 downto 16):

```
lui Rt, Imm-upper
```

In order to load the lower half part of the register, an *or immediate* (`ori`) instruction can be used that loads the lower half (16-bit) part of the register leaving the upper half part unchanged.

```
ori Rt, Rt, Imm-lower
```
[35]

Therefore, for the application of the *k* test patterns to the subtracter the following self-test routine can be applied.

```
test-subtracter: andi R4, R4, 0
                 lui R2, xtest-1-upper
                 ori R2, R2, xtest-1-lower
                 lui R3, ytest-1-upper
                 ori R3, R3, ytest-1-lower
                 sub R1, R2, R3
                 sw R1, rstart(R4)
                 addi R4, R4, 4

                 lui R2, xtest-2-upper
                 ori R2, R2, xtest-2-lower
                 lui R3, ytest-2-upper
                 ori R3, R3, ytest-2-lower
                 sub R1, R2, R3
                 sw R1, rstart(R4)
                 addi R4, R4, 4
                 ...
                 lui R2, xtest-k-upper
                 ori R2, R2, xtest-k-lower
                 lui R3, ytest-k-upper
                 ori R3, R3, ytest-k-lower
                 sub R1, R2, R3
                 sw R1, rstart(R4)
                 addi R4, R4, 4
```

This self-test code consists of unrolled segments of code, each of which applies one test pattern to the subtracter component. Registers `R2` and `R3` are loaded with the 32-bit values in two instructions, the `lui` instruction that loads the upper half of the register and the `ori` (or immediate) which loads the lower half of the register while leaving the upper half unchanged. The test pattern is applied by the `sub` instruction while the test responses of the subtracter are collected in an array of memory words starting at address

[35] The sequence of `lui` and `ori` instructions is combined as an assembler macro (or pseudoinstruction) called `li` (load immediate) which loads a register with a full 32-bit immediate value using `lui` and `ori`. We have decided not to include macros (or pseudo-instructions) in our code examples, so that the reader can quickly calculate the total number of words of the example, assuming one word per instruction. Macros, usually are equivalent to more than one word each.

`rstart` as in the previous case of loop-based application and register `R4` is used as an index of the test responses array.

Figure 5-16 presents in a visual way the application of ATPG-generated test patterns using immediate addressing mode instructions. Four instructions are required to load the two input registers with the test patterns not using any part of the data memory but instead of the instructions themselves (two instructions for each register – the `lui` and `ori`). The test patterns are then applied by the subtract instruction to the subtracter component as the code above shows.

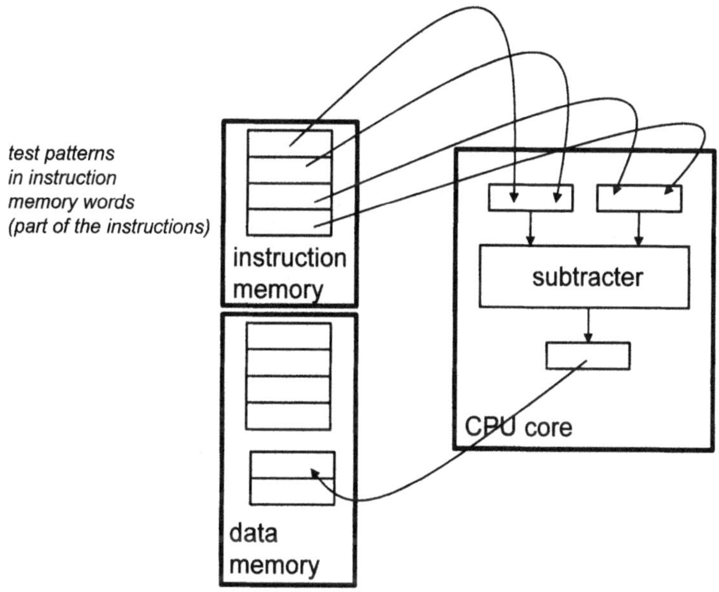

Figure 5-16: ATPG test patterns application with immediate instructions.

The code using immediate addressing mode consists of $7k$ words (seven words for each of the k test patterns). Assuming again a one cycle execution for each instruction apart from memory reads and writes which take 2 cycles; the total execution time of the code is equal to about $8k$ clock cycles.

The size of the last self-test routine for immediate addressing application of the ATPG-based test patterns is due to the fact that the uncorrelated k test patterns for the subtracter can't be applied in any loop-based manner (such as in the case of application from the memory shown before). On the other side, this unrolled coding style leads to a smaller routine execution time: $8k$ compared with the $10k$ time of the loop-based routine we saw before.

Table 5-2 summarizes the characteristics of code size and execution time for the two different cases of application of ATPG-based test patterns we just studied. Moreover, the table shows the total test application time when the two routines are applied using a 50 MHz tester and the operation frequency of the processor is 200 MHz (see the calculations given in Section 5.2.3). The number of test responses is *k*.

Processor frequency – 200 MHz
Tester frequency – 50 MHz

Coding style	Words	Responses	Cycles	Test Application Time
ATPG-based loop from memory	$2k+8$	k	$10k$	$0.11\,k+0.16$ μsec
ATPG-based with immediate	$7k$	k	$8k$	$0.20\,k$ μsec

Table 5-2: ATPG-based self-test routines test application times (case 1).

We can see in Table 5-2 that in this case where the processor is faster from the tester to a reasonable level (processor is 4 times faster than the tester), the loop-based application of ATPG-based patterns from data memory is about two times faster in test application time compared to the immediate addressing mode application of the same *k* test patterns.

If, on the other side, the tester is ten times slower than the processor chip (consider for example a tester of 100 MHz frequency used for a 1 GHz processor) then the difference between the two approaches becomes more significant, as Table 5-3 shows.

Processor frequency – 1 GHz
Tester frequency – 100 MHz

Coding style	Words	Responses	Cycles	Test Application Time
ATPG-based loop from memory	$2k+8$	k	$10k$	$0.04\,k + 0.08$ μsec
ATPG-based with immediate	$7k$	k	$8k$	$0.088\,k$ μsec

Table 5-3: ATPG-based self-test routines test application times (case 2).

It is also useful to make another remark regarding the ATPG-based approach for component self-test routines development. The remark is that when the value of *k* is small (this is the case for some relatively simple processor components) this approach (either in the loop format from memory or in the immediate format) gives a very good solution to software-based self-testing because the test routine size and its execution time will both be very small and the obtained fault coverage will be very high.

Unfortunately, for the most important components of a processor it is not always possible to derive a small test set using an ATPG. Therefore, when the number *k* is large, either the size of the self-test routine is excessively large (this is the case when the immediate addressing mode is used) or the execution time of the program of excessively high (this is the case in both cases of loop-based approach and the immediate addressing mode approach).

We have to note at this point that the clock cycles that the load word and store word instructions need (the CPI – clocks per instruction) has an impact on the exact values of the previous Tables which are only indicative. Also, the self-test routines execution time is affected by the existence or not of a pipeline structure.

5.7.2 Self-test routine development: pseudorandom

Pseudorandom testing is another popular approach in digital systems testing, mainly because of its simplicity in test generation (actually, there is no test generation phase! in the sense that ATPG defines it).

When pseudorandom testing is applied to a digital circuit, a long sequence of test patterns is usually applied to the circuit's primary inputs. In hardware-based self-testing the pseudorandom test patterns are generated by hardware state machines specially synthesized for this purpose[36]. The most common hardware pseudorandom pattern generators are the Linear Feedback Shift Registers (LFSRs), the Cellular Automata (CA) and the pseudorandom pattern generators based on arithmetic circuits like adders, subtracters, multipliers. All these three types of circuits along with several modified versions presented in the literature have been proved to have excellent properties in generating sequences of random-like patterns.

LFSR-based pseudorandom testing is the most popular pseudorandom technique. The quality of the pseudorandom sequences depends on the initial value loaded into the register (called *seed*) and the *polynomial* that the internal connections of the LFSR realize.

LFSR-based pseudorandom self-testing can be also implemented using embedded self-test routines. Self-test routines actually *emulate* the generation of pseudorandom pattern sequences by realizing the polynomial evaluation in software instead of hardware. There are two different ways that an LFSR-emulating pseudorandom self-testing approach can be applied to a processor core to test its modules:

- development of a *separate* pseudorandom self-test routine for each component that applies the generated test patterns directly to the component;
- development of a *main, programmable* LFSR-emulation self-test routine that will generate pseudorandom sequences after being

[36] The hardware-based self-testing scheme where test patterns are applied from the memory does not apply to pseudorandom testing because of the large number of test patterns.

called by component self-test routines and apply them to components

The advantage of the first approach of separate routines for pseudorandom patterns generation is that the interaction with the memory is very limited and patterns can be directly applied to the component under test without having been put in the memory before. The drawback of this approach is that the self-test code size is larger because a separate routine (or even a set of routines) is developed for each component around the basic processor instructions that apply the tests to the component operations.

An example of the code in this first approach of pseudorandom based self-testing is given below again for testing the subtracter component of the processor. We assume that an LFSR is emulated by processor instructions. Other pseudorandom patterns generators schemes can be employed as well (like cellular automata or arithmetic circuits) but we show the LFSR-based scheme in our example code since this is the most classical scheme in hardware-based pseudorandom self-testing and has also been applied to software-based pseudorandom self-testing for processor cores.

```
test-subtracter: sub R11, R11, R11
                 ori R8, R0, max-patterns
                 lw R9, polynomial-mask-x
                 lw R10, polynomial-mask-y
                 lw R2, seed-x
                 lw R3, seed-y
next-sub:        sub R1, R2, R3
                 sw R1, rstart(R11)
                 addi R11, R11, 4
                 subi R8, R8, 1
                 beq R8, R0, exit
                 ori R4, R2, 0
                 andi R4, R4, 1
                 srl R5, R2, 1
                 beq R4, R0, complete-x
                 xor R5, R5, R9
complete-x:      andi R2, R5, FFFF
                 ori R4, R3, 0
                 andi R4, R4, 1
                 srl R5, R3, 1
                 beq R4, R0, complete-y
                 xor R5, R5, R10
complete-y:      andi R3, R5, FFFF

                 j next-sub
exit:
```

In the above self-test code, registers `R2` and `R3` are the pseudorandom operands that will be applied to the subtracter at each iteration of the loop. The result is collected in register `R1` and stored to the responses array after subtraction has been performed. Register `R8` counts the number of pseudorandom patterns applied and initially contains the maximum number (number of iterations of the main loop). Registers `R9`, `R10` contain the masks that implement the characteristic polynomial of the software-emulated LFSRs. Registers `R2` and `R3` are initially loaded with the seeds. We assume in our example that a different LFSR is implemented for each of the X and Y operands (`R2` and `R3` registers). Registers `R4` and `R5` are used to implement the algorithm of the LFSR next value calculation (extract the rightmost bit; look if it is a 1; right shift the previous value; xor with the mask if rightmost bit is 1). Register `R11` is the index to the test responses array.

We note that the routine above applies pseudorandom test patterns to a subtracter using its basic instruction (`sub`). In order to test each component and each operation of every component, a separate routine must be designed and applied to the component. The routine above consists of 24 words. Alternatively, more than one instructions can be applied to the processor for each new pseudorandom pattern (or pair of X, Y patterns) generated. This means that after the `sub` instruction, another one will be inserted and applied to the same (or other) component and response will be transferred out of the processor. Each iteration of the basic loop above that generates and applies an new pattern takes 16, 17 or 18 cycles (because of the taken or untaken branches of the above code depending on whether the rightmost bit of the current LFSRs value is 0 or 1).

The advantage of the second approach of a main, programmable routine for pseudorandom patterns generator is that it avoids many copies of instructions needed for each of the separate routines for each component. The drawback of this approach is that it requires several calls to other routines in the program, which will increase the overall duration of the self-test program execution. Therefore, it is again a decision regarding the tradeoff between longer test application time and larger self-test routine.

The outline of a self-test routine according to the second approach is the following:

```
test-component: ori R8, R8, max-patterns
                # initialize LFSR routine (seed, mask)

next-pattern:   # call LFSR routine for first operand
                # call LFSR routine for second operand
                # apply the target instruction
                sw R1, rstart(R11)
                addi R11, R11, 4

                subi R8, R8, 1
                beq R8, R0, exit

                j next-pattern
exit:
```

In the code above, we assume that the test response is collected in `R1` and then stored to the responses array. The LFSR routine can be called as many times as necessary for the specific operation being tested (usually two times for two-operand operations, so that in the first time the one operand's next random value is generated and in the second time the other operand's random value is generated).

We mention at this point that all code examples given in this Chapter using the MIPS instruction set may be differently applied to other processor architectures and of course may also be differently applied in MIPS-like processors depending on the coding style that an assembly programmer prefers and the specific requirements of the application. Several changes can be done to the code examples given in this book. The examples we present are only indicative, and their purpose is to give an idea for how the related approaches may be applied, using as a demonstration vehicle a popular instruction set architecture of a successful processor.

5.7.3 Self-test routine development: pre-computed tests

A third component test development approach is based on the use of regular sets of known, pre-computed test patterns for its key components. Such pre-computer test patterns are very useful and effective when the netlist of the component (or the entire processor) is not available for ATPG. Sets of known, pre-computed test patterns can be stored in a components test patterns library and accordingly applied to processor's components.

In the literature, there has been proposed several such test sets which are either characterized by a small number of test patterns or by a regularity and correlation between the test patterns they consist of. In the former case, such pre-computed test sets were basically developed for external testing and this is why the test set size is small. In the latter case, these test sets were developed in such a way so that efficient hardware built-in self-test generators can be designed for their on-chip production and delivery to the components under test. Such test patterns are deterministic and not pseudorandom and have been specially designed to obtain high fault coverage for common architectures of several components. Some of these test patterns sets have been shown to obtain very high structural fault coverage for any word length of the components and any of its different internal architectures[37]. In that sense, the regular, deterministic test sets are *generic* and when applied to a processor component, no fault simulation is necessary to prove their effectiveness. Of course, the exact fault coverage they obtain on a particular circuit can only be evaluated if a gate-level netlist is available, but even with no fault simulation these test sets are known to provide very high structural fault coverage.

Processor components which can be effectively tested with a set of deterministic, regular test patterns are:

- the arithmetic and logic components: adders, subtracters, ALUs, multipliers, dividers, shifters, barrel shifters, comparators, incrementers, etc; (*computational functional components*) [20], [53], [54], [55], [115], [162];

[37] See, for example, the works on multipliers testing [53], [54], [95], detailed in a few pages.

- the registers and register files (*storage functional components*) [98];
- the multiplexers of the data busses (*interconnect functional components*).

The main advantages of the use of regular, pre-computed, deterministic test patterns are the following:

- *component architecture independence*: this applies when the same test set is a priori known to be effective (at least to some sufficient level) for different architectures of the component;
- *small test set size*: regular deterministic test sets consist of a small number of test patterns;
- *ease of on-chip generation*: this is true not only for hardware based self-testing but also for software-based self-testing as we see in the following paragraphs

Regular, deterministic test sets consist of test patterns that have a relation between them, which makes them easily on-chip generated (either by hardware or by software). Each test pattern of the test set can be derived from the previous one by a simple operation such as:

- an arithmetic operation (addition, subtraction, multiplication);
- a shift operation
- a logic operation
- combination of the three above

One may think that pseudorandom tests are generated in a similar way (LFSR emulation consists of shifts and exclusive-or operations). The difference between the two approaches is that pseudorandom testing requires the generation of a large number of test patterns, while regular, deterministic relies on a small test set.

Regular, pre-computed, deterministic test sets combine the positive properties of ATPG-based and pseudorandom based test development for processor components. Therefore, in cases when ATPG-based test development is not able to obtain high fault coverage with very few test vectors, and also in cases when pseudorandom test development is not able to obtain a high fault coverage within a reasonable amount of time (clock cycles), regular pre-computed sets of test patterns is the solution. They reach high fault coverage levels with a number of test patterns (and thus clock cycles) much less than pseudorandom testing and a bit more that ATPG-based test development. On-chip generation of regular, deterministic test sets is as easy as pseudorandom test development while the self-test program size is reasonably small.

As a key example, we mention the case of multipliers testing [53], [54] where it was proven that a regular test set of 256 test patterns can obtain a fault coverage of more than 99% for different types of array multipliers (array multipliers, tree multipliers, Booth encoded or not) and for any word length of the multiplier. This test set of 256 test patterns (which was later shown to be equally sufficient when test patterns were reduced to 225 and 100 respectively [52], [93], is a little larger than the test set that an ATPG can produce for an array or tree multiplier and much smaller that the test set that is necessary to obtain such a high fault coverage with pseudorandom testing (around 200 test patterns can reach a 95% fault coverage, while a 99% fault coverage needs more that 1000 or even 2000 test patterns, depending on the multiplier structure and gate-level details. Moreover, the regular test set can be very easily generated on-chip by a dedicated, counter-based circuit (equally easy as an LFSR-based hardware self-testing approach).

In the case of software-based self-testing, such regular, deterministic test sets can be also easily generated by compact self-test routines.

Table 5-4 summarizes the characteristics of the four different component-level test development techniques, where self-testing is performed using embedded software routines.

Approach	Fault coverage	Self-test code size	Test application time
ATPG loop from memory	High	Small	Short
ATPG with immediate	High	Large	Short
Pseudorandom	Medium	Small	Long
Regular Deterministic	High	Small	Short

Table 5-4: Characteristics of component self-test routines development.

5.7.4 Self-test routine development: style selection

Now, that we have analyzed and discussed the different test development approaches for the processor components, the question to be answered is: *which one of the approaches is selected for a particular component of a processor?* The answer may seem straightforward after the long analysis, but it is always useful to summarize.

We select the ATPG-based, self-test routine development approach for processor components when:

- the component is a small, combinational or sequential component and an ATPG can generate a small number (some tens) of test patterns;
- a gate-level netlist of the component is available or can be obtained;
- constraints extraction can be performed for the component, so that ATPG is guided by the constraints and is able to generate test patterns that can be applied only with processor instructions;
- there are no pre-computed test sets known for this type of processor component;

In ATPG-based test development, high fault coverage is guaranteed (provided that the ATPG tool succeeds to obtain high fault coverage).

If the question is which of the two ATPG-based code styles to select (fetching test patterns from memory or using the immediate addressing format), this decision depends on the actual number of test patters. If they are very few then immediate addressing mode is simpler to be applied (simple writing of the assembly code).

We select the pseudorandom-based, self-test routine development approach for processor components, when:

- the component is expected or known to be pseudorandom pattern testable and not resistant (even if the gate level netlist of it is not available for fault simulation);
- a gate-level netlist of the component is not available or can't be obtained;
- there are no pre-computed test sets known for this type of processor component;

The actual fault coverage obtained in the pseudorandom case, can only be calculated if a gate-level netlist is available, but the test development phase (development of the self-test routines) can be done without any gate-level information available.

We select the regular, pre-computed pattern based, self-test routine development approach for processor components, when:

- there are known, effective, pre-computed test sets for this type of component (even if the gate level netlist of it is not available);
- a gate-level netlist of the component is not available or can't be obtained by synthesis;

5.8 Test development for processor components

After having finished the discussion on component-level test development for processor components and the comparison between them, we elaborate further on the specifics of the different classes of processor components (functional, control and hidden).

5.8.1 Test development for functional components

The analysis of the different component-level self-test routine development approaches, given in the previous sections, mainly focused on the functional components of the processor (classified in the computational, storage and interconnect sub-classes). For the functional components, all three self-test routine development styles (ATPG-based, pseudorandom, pre-computed/deterministic) can be applied provided that their requirements for test development are met.

The most important components of a processor belong to the functional category and, therefore, self-test routines development for them can completely be done using one of the self-test routine development approaches presented so far.

The experiment results presented in the next Chapter support our argument that the most important components of the processor (i.e. the larger and more easily testable ones) are actually the functional components. Software-based self-test has been applied to several publicly available processor models of different architectures, assembly languages and word lengths. The experimental results prove our claim that software-based self-testing can be an excellent, low-cost self-testing methodology which obtains very high fault coverage levels with small self-test programs.

5.8.2 Test development for control components

The second category of processor components, the control components, has also a central role in the correct execution of the user programs that run on the processor and thus must be considered for software-based self-test generation, right after the most important components: functional.

Control components operation is related to:

- instruction fetching and decoding;
- addressing modes implementation and memory access (instruction and data memory);
- internal and external busses interfacing, etc;

In most cases, control components of the processor are relatively small in size and therefore their contribution to the overall processor fault coverage is relatively low.

Let us consider the case of the control unit of a processor, that produces the control signals for the several computational, storage and interconnect functional components. Such control units are in many cases implemented as finite state machines, and can be realized in many different ways (micro-programmed units, hardwired units, etc).

Testing of control units can be based on:

- *functional software-based self-test*: the control unit (state machine) can be tested by ensuring it passes through all of its states (if all of them are known) and it passes through all normal transitions between states (this is an exhaustive testing approach, which is infeasible for large FSM designs);
- *scan-based,hardware-based self-test*: if structured DfT modifications are allowed, this is an efficient approach in many cases, because DfT changes in the control unit most likely do not have an impact on the critical path of the processor and thus can be easily adopted by the chip designers;

Moreover, it has been observed (also noted in the experimental results of the next Chapter) that, while self-test routines are being developed for the functional components of the processor, they, in parallel, reach sufficient fault coverage levels for the control units as well. This is particularly true, if a relatively rich variety of processor instructions are used in the self-test routines for the functional components. Therefore, processor control components may be tested to some extend, as a side effect of the testing of the processor's functional components.

If a control unit component must be specifically targeted by a self-test routine, then this routine should include an exhaustive excitation of the component's operations. For example, all the different instructions and all different addressing modes of the processor must be applied if we want to efficiently test an instruction decoding component. Such self-test routines can be based on already existing routines previously developed for design verification of the processor by designers. In design verification of the processor, it is very common that a self-test program executes all different processor instructions to prove that the processor works correctly in all these cases. Such a routine (or set of routines) can be re-used to effectively test the processor's control unit.

The experimental results of the next Chapter prove both the arguments that control components are small in size and that they can be tested to some sufficient level while self-test routines dedicated to functional components are applied to the processor.

5.8.3 Test development for hidden components

The third category of processor components, according to our classification scheme, consists of the hidden components of the processor. The role of the hidden components is to improve the processor's performance, although the assembly language programmer is not aware of their existence and their actual structure is not visible.

The most important types of hidden components related to performance improvement are the components that implement pipelining. Pipelining has been the most important and successful performance improving technique for processors during the last decades. A pipeline mechanism implemented in a processor consists usually of:

- a set of large multi-bit *pipeline registers* for the transfer of instructions and control signals down the pipeline stages; this way each instruction carries with it all necessary information to continue execution although new instructions have already been issued in previous pipeline stages;
- a set of large *multiplexers* for the selection of different sources of data when *forwarding* is implemented; this way pipeline stalls are avoided due to data dependencies between instructions;
- a *control logic* at each pipeline stage, which is usually very small; for example pipeline logic detects the existence of data hazards and activates forwarding or stalls, etc;

In the context of low-cost, software-based self-testing for processors, it is not completely suitable for self-test program generation to target hidden components such as the components of the pipeline structure of a processor. Aiming to detect, with small and fast self-test programs, faults in components that are not visible and their existence can't be implied from the known information, may either be infeasible or very costly. Therefore, self-test routines development for such components is not of high priority for software-based self-testing. All that can be done is discussed in the following paragraphs.

Since pipeline existence (or at least its exact implementation and details) is not known to the assembly language programmer, the components related to it can't be targeted directly by software-based self-test routines. On the contrary, they can be indirectly tested when other components of the processor are being tested. In many cases, this indirect testing can be quite effective.

Let us elaborate further on each of the three pipeline components above:

- *Pipeline registers*: these registers are tested while the other components of the processor are being tested; this is because the instructions of the self-test routines pass through all the stages of the pipeline applying a variety of test patterns to them. Faults in the registers are excited and also easily propagated to the primary outputs of the processor since every instruction carries all necessary bits all the way down the pipeline final stage. In this sense, the pipeline registers can be considered easy-to-test components. Our experimental results show that this statement is valid at least for our benchmark processors.
- *Pipeline multiplexers*: these multiplexers can be easily tested as well, if the self-test routines for other components of the processor make sure that the different forwarding paths of the pipeline are excited; this can only be done if RTL information for the actual implementation of the pipeline structure is available for the processor.
- *Pipeline control*: it consists of a relatively small number of gates for each pipeline stage (mainly a set of small comparators that compare parts of the instructions in consecutive pipeline stages to identify necessary forwarding and potential pipeline stalls), and thus does not require special test development and is expected to be tested to some extend when the pipeline registers and pipeline multiplexers are tested.

Pipeline multiplexers are large multi-bit multiplexers that implement the forwarding operation in pipelines, work together with the relatively small comparators of the forwarding unit. In case that the self-test programs developed previously for the functional components of the processor do not contain appropriate instruction sequences that excite all paths through the multiplexers and comparators of the pipeline login, special routines can be developed. The development of such routines can only be done if some information for the actual implementation of the pipeline structure of the processor is available, so that all different forwarding paths are used and therefore all multiplexers and comparators are sufficiently tested.

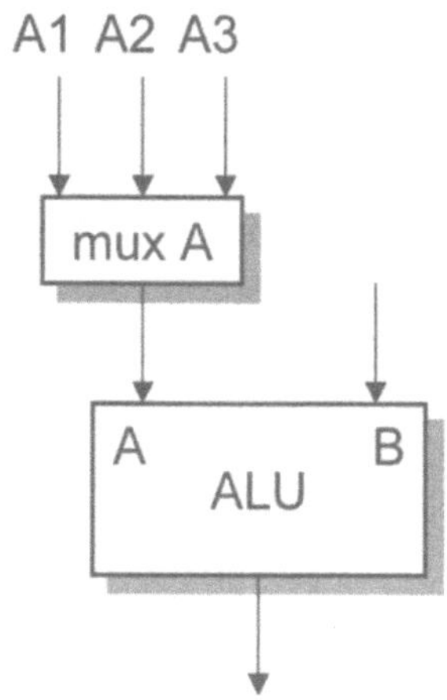

Figure 5-17: Forwarding logic multiplexers testing.

For example, Figure 5-17 shows part of the pipeline logic where a 3-input multiplexer is used at the A input of the ALU to select from three different sources. Signals A1, A2, A3 are connected to the ALU input A depending on the forwarding path activated in the processor's pipeline (if such a path is activated at this moment). A successful self-test program generation process must guarantee (if RTL or structural information of the processor is available) that all paths via the multiplexer (A1 to A, A2 to A, A3 to A) are activated by corresponding instructions sequences so that the multiplexer component is sufficiently tested.

As we see in the experimental results of the next Chapter, pipeline structures of the processors are relatively easily tested while self-test routines are being applied to the other processor components. Of course, we point out that the publicly available benchmark processors we use may have a relatively simple pipeline structure compared with high-end modern microprocessors or embedded processor. Unfortunately, the unavailability of such complex commercial processor models makes the evaluation of software-based self-testing on them difficult. Fortunately, software-based self-testing is attracting the interest of major processor manufacturers and it is likely that results will become available in the near future. This way, the potential limitations (if any) of software-based self-testing may be revealed.

Moreover, software-based self-testing must be evaluated for other performance-improving mechanisms such as the branch prediction units of processors. In such cases, due to the inherent self-correcting nature of these mechanisms, faults inside them are not observable at all to processor outputs, but only performance degradation is expected. Performance measuring mechanisms can be possibly used for the detection of faults in such units and research in this topic is expected to gain importance in the near future.

5.9 Test responses compaction in software-based self-testing

So far, we concentrated on the test pattern generation and delivery part of software-based self-testing, while we assumed that all test responses of the components under test were collected in a responses array in the processor's data memory. Under this assumption, each test response can be separately compared with the expected fault-free response, either by the processor itself or by external equipment.

If a compacted self-test signature must be collected for all test patterns applied to the processor, or a signature for each component under test, then a special response compaction routine must be developed either for each of the components (if a per component signature is needed) or for the processor in total.

Compaction of self-test responses can be performed in two different ways:

- *two-step compaction*: component test responses are stored in a data memory array and are then separately processed by a response compaction routine that produces a single signature for the entire processor or a single signature for each of the targeted components;
- *one-step compaction*: component test responses are compacted "on-the-fly" while they are generated; part of the components self-test routines is specially designed to compact the new test response with the previous value of the self-test signature;

In two-step compaction, the component self-test routines may be smaller since they do not include code for response compaction and simply store a response to data memory. As we know, smaller self-test routines require smaller download time from the external low-cost testers. The disadvantage of two-step compaction scheme is that is has a very intensive communication with the data memory (one write for each new component test response), and this may impact the overall test application time (routine execution) due to the long response time of memory. Moreover, two-step response compaction occupies a significant amount of memory where the test responses must be collected.

On the other side, one-step compaction of test responses leads to larger self-test routines for the processor component and thus larger download time, but they have a reduced interaction (if any) with data memory and are therefore, most probably, faster routines than those with two-step compaction.

Figure 5-18 depicts two-step response compaction while Figure 5-19 depicts one-step response compaction.

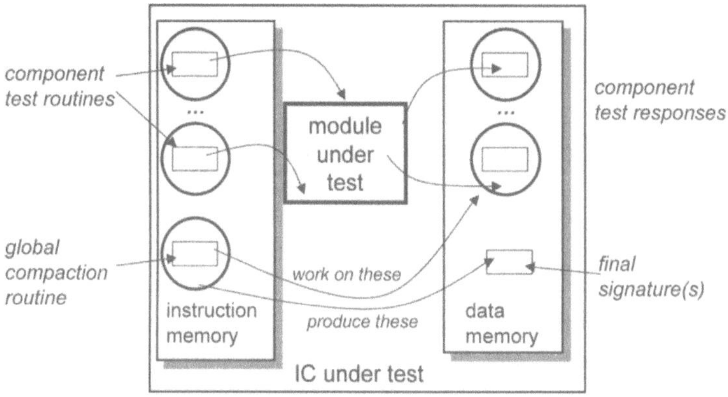

Figure 5-18: Two-step response compaction.

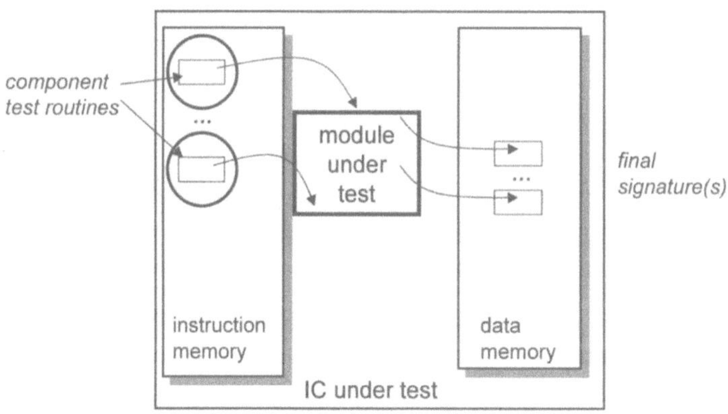

Figure 5-19: One-step response compaction.

The compaction schemes that can be used in software-based self-testing are the same as in hardware-based self-testing since the same properties are seeked. The primary concern in test responses compaction is always the fault coverage loss due to fault masking and aliasing. A smaller number of collected signatures leads to higher probability of aliasing, while more signatures reduce this probability and related fault coverage loss.

In pseudorandom software-based self-testing, aliasing probability is theoretically smaller than in ATPG-based or regular patterns based, simply because the number of test patterns applied is much larger in pseudorandom testing. The low aliasing properties of compaction schemes based either on Multiple Input Signature Registers (MISRs) or in Accumulator-Based Compaction (ABC) scheme, are valid when the number of test patterns are very large. On the contrary, when a few deterministic (ATPG or pre-computed) test patterns are applied, aliasing due to a compaction scheme strongly depends on the exact patterns that are applied and therefore no general argument can be valid. Experimental results are required to determine if a compaction scheme is effective or not for a particular processor component.

5.10 Optimization of self-test routines

Software-based self-testing is usually applied at the component level, i.e. separate self-test routines are developed and subsequently applied to each one of the targeted processor components. Not all the components are targeted, since a sufficient fault coverage may be reached when self-test routines are developed only for a subset of them (usually all the functional components and some of the control components). Under this model, if components C_1, C_2, C_3 and C_4 have been targeted and self-test routines P_1, P_2, P_3 and P_4 have been developed for them with sizes (in bytes or words) S_1, S_2, S_3 and S_4 then the total self-test program for the processor has a total size equal to $S_1+S_2+S_3+S_4$. If the execution time of each of the four routines is T_1, T_2, T_3 and T_4, respectively, then the total execution time for the program will be approximately equal to $T_1+T_2+T_3+T_4$ (some extra time will be necessary for invoking the four routines from a main routine).

In many cases, optimization of a self-test program for a processor is necessary to make it more efficient in terms of size (smaller program) or execution time (faster program). This optimization is necessary in order to:

- reduce the self-test program download time from the low-cost external tester;
- reduce the self-test program execution time;

As a consequence, the total test application time (download + execution) will be reduced for each processor being tested. Of course, this last phase, of self-test code optimization is optional and may be skipped when the initial self-test program size and execution time are sufficiently small for the particular processor and application, and no extra engineering effort is necessary (or can be afforded) to optimize it further. Self-test program optimization is the fourth step of software-based self-testing (Phase D of Figure 5-7.

In the two next sections we discuss two different self-test code optimization approaches that may be applied separately or together. Other techniques may be applied also.

5.10.1 "Chained" component testing

This self-test code optimization technique is based on the successive testing of processor components where the test response of one of them is used as a test pattern for the next component. Therefore, components are put in a virtual "chain" and are tested one after the other. If, for example, three processor components C1, C2 and C3 are "chained" together in this order, then an optimized self-test program repeats several cycles where a test pattern is applied to C1, its test response (stored in a register) is applied as a test pattern for C2, its test response (stored again in a register) is applied as a test pattern to C3 and the test response of C3 is transferred out of the processor as the test response of the "chained" testing of the three components.

This type of self-test code optimization has the potential to develop very compact self-test routines that test all grouped components to a high level of fault coverage. Its successful application depends on the following factors.

- The function that a component performs must be able to give a sufficient variety of responses that can be used as test patterns for the subsequent components in the chain.
- Errors at the outputs of the first components of a chain (due to faults inside them) must be propagated to processor primary outputs after they pass through subsequent components of the chain. This must be true throughout the entire chain of components.

Self-test code size and execution time reduction is almost guaranteed by this technique although in some case there may be some compromise in the fault coverage level because of observability problems, due to the second factor given above.

The following simple example gives an idea of this optimization technique considering that two components: an adder and a shifter are tested together in a chained fashion. Figure 5-20 depicts this idea. Test instruction 1 applies a test pattern to the adder. The adder test response is captured in a register. The content of this register is used by test instruction 2 as a test pattern for the shifter component. The shifter output is the combined test response of the two components.

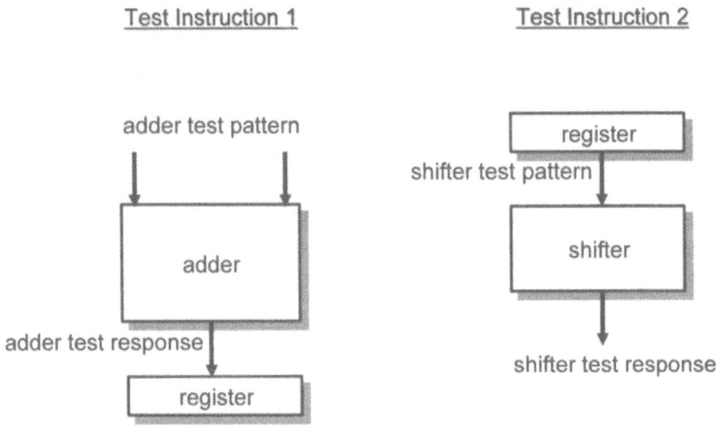

Figure 5-20: "Chained" testing of processor components.

In this example, successful application of the chained testing of the two components requires the following:

- A sufficient test set for the shifter can be generated at the outputs of the adder. In other words, appropriate inputs must be supplied to the adder so that they test the adder itself and also provide a sufficient test set for the shifter.
- The shifter does not mask the propagation of error at the adder outputs (because of faults in the adder) towards primary outputs of the processor.

Let us consider that, if separate self-test routines are developed for the two components, adder and shifter, each of them consists of a basic loop that applies a set of test patterns to each component. Let us assume also, that the test set for the adder applies 70 test patterns to it (this can be the case for a carry lookahead adder) and that the test set for the shifter applies 50 test patterns to it. Also, let us consider that the basic loop of the self-test routine for the adder executes each iteration in 30 clock cycles, and that the basic loop of the self-test routine for the shifter also executes each iteration in 30 clock cycles. Therefore, a self-test program that uses these two routines one after the other will execute in approximately $70 \times 30 + 50 \times 30 = 3{,}600$ clock cycles. Applying the "chained" testing optimization technique will most probably lead to a combined loop that applies to the adder and shifter a total of 80 test patterns and executes each iteration in 31 clock cycles (one more instruction is added to the loop). The total number of test patterns is larger since it may be necessary to "expand" the set of adder tests so that their responses produce a sufficient test set for the shifter which is tested subsequently. Moreover the larger loop execution time for each execution (31 instead of 30) is due to the fact that an extra instruction is added to it for the application of the test pattern to the shifter. In the combined, "chained" routine the total execution time of the combined loop will be $80 \times 31 = 2{,}480$ clock cycles. The numbers used in this small analysis are picked to show the situation in which the optimization technique can lead to a more efficient self-test code. There may be, of course, situations where the optimized code is not better than the original. For example, this may be the case when the total iterations of the combined loop are too many. This may be caused by the inability to reasonably "expand" the adder test set so that a sufficient test set for the shifter is produced at the adder outputs. If, in our simple adder/shifter example the total number of combined loop iterations is 120 instead of 80 then the number of clock cycles for the new loop will be $120 \times 31 = 3{,}720$ which is larger from the original back-to-back execution of the two component routines.

The following pseudocodes show how "chained" component's testing can be applied. First, two separate self-test routines for component's C1 and C2 are given. Then, the combined routine for the "chained" testing of the components is shown. In this example we assume that the components are both originally tested with ATPG-based test patterns applied in a loop from data memory.

```
atpg-loop-C1:
      load register(s) with next test pattern for C1
      apply instruction IC1
      store result(s) to memory
      if applied-patterns < K1
          repeat atpg-loop-C1

atpg-loop-C2:
      load register(s) with next test pattern for C2
      apply instruction IC2
      store result(s) to memory
      if applied-patterns < K2
          repeat atpg-loop-C2

atpg-loop-C1-C2:
      load register(s) with next test pattern for C1
      apply instruction IC1
      apply instruction IC2
      store result(s) to memory
      if applied-patterns < max(K1, K2)+m
          repeat atpg-loop-C1-C2
```

Although we use an ATPG-based self-test routine style in the pseudocode where patterns come from data memory, any of the other self-test routine styles can be applied in "chained" components testing.

Chained testing of processor components can lead to self-test code size and execution time reduction because:

- Only one self-test routine in necessary for all components in the chain. The routine may be smaller than the sum of all the separate routines.
- The single self-test routine will execute faster than the combination of the separate routines because it will contain only one slightly slower loop instead of several ones, and also it will have a much smaller number of interactions with the memory system (to store test responses). Only the combined test response of the final component in the chain is stored to memory.
- The combined self-test routine will produce a much smaller number of self-test responses into processor memory and therefore the uploading of these responses towards low-cost tester memory will be shorter and total test application time will be reduced.

In summary, "chained" testing of processor components has several advantages over individual component-level self-testing. It requires some more sophisticated analysis to be done (on the feasibility of "chained" testing for a particular processor) but it leads to significantly smaller self-test programs with less loops, less memory interaction and less test responses. Therefore, a smaller test application time of the device is gained.

5.10.2 "Parallel" component testing

According to this second alternative for the optimization of the self-test routines, the same set of test patterns is applied to more than one component and therefore, the application of separate self-test routines to each of them is avoided. We call this technique "parallel" testing of the processor components, to denote that the components are tested with the same set of patterns. Figure 5-21 shows "parallel" testing of processor components for a pair of an adder and a subtracter component. The same test pattern is applied to both of them with separate instructions (addition instruction for the adder and subtraction for the subtracter).

This second optimization technique does not have the drawback of fault masking or fault blocking that the "chained" testing technique has. In "parallel" testing of processor components, each test response from the components is stored to the memory and not used as a test pattern for another component.

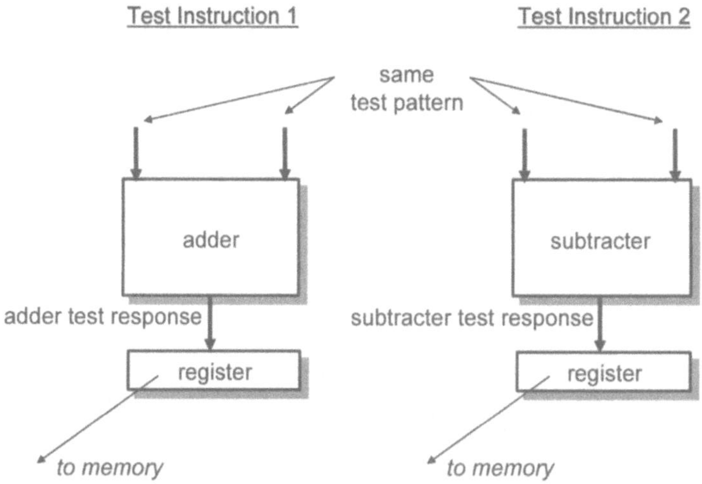

Figure 5-21: "Parallel" testing of processor components.

The benefit of the "parallel" testing of processor components is that the loops that generate the component tests are combined together to form a new, global loop for the components that are now tested in "parallel". There is no need for a separate generation of the next test pattern for each of the components. Therefore, the self-test code size is significantly reduced and also the total execution time for the new loop is much smaller than the combined execution time of the separate loops. The number of test responses is not reduced as in the case of "chained" testing of components but may rather be increased. This can happen when a smaller test set for a component is expanded to a larger one that contains the small one (or obtains fault coverage similar or same to the small one).

5.11 Software-based self-testing automation

Every testing and self-testing methodology requires a set of supporting algorithms for its application and also requires a successful implementation of EDA tools that realize these algorithms efficiently. Manual test development, although useful and very efficient in small circuit sizes, can't be easily applied to large circuits.

Software-based self-testing of embedded processor cores requires corresponding automatic tools for the support of the generation of self-testing routines to test the processor components separately and the processor in total. Figure 5-22 shows a graph with parts of the software-based self-testing flow for a processor that can be automated.

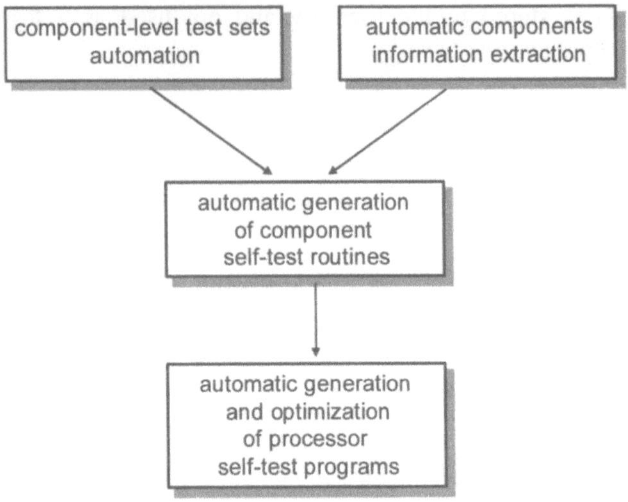

Figure 5-22: Software-based self-testing automation.

A first part that can be automated is the component test development part, in which a set of test patterns that can be applied by processor instructions is derived for each of the components that the processor consists of. In some cases, this part has been implemented using constraints extraction for the components and feeding the extracted constraints to an ATPG. If successful, this process leads to a set of ATPG-generated test patterns for the target processor component which can be applied to it using processor instructions. Several attempts to this direction in the open literature show the importance of this path (for a recent, excellent analysis of this problem, the reader may refer to [31]). Automatic (as opposed to manual) extraction of instruction set imposed constraints for components testing will lead to significant reduction of the test generation (and self-test routines generation) time. The advantage of this approach is that, if successful, can be universally applied to different processors and different instruction set architectures.

A second part of software-based self-testing that can be automated is the part of extraction of the information required to develop the self-test routines. This information has been discussed earlier in this Chapter and consists of:

- the set of processor components from its RTL description;
- the set of operations that each component of the processor executes (as instructed by processor instructions);
- the set of different instructions of the processor that excite a particular operation of a component;

- the controllability and observability of processor registers and processor components.

The third part of software-based self-testing than can be automated is the development of the component (and processor) self-test routines themselves. In particular, a tool for the automatic generation of self-test routines for processor testing should be supplied with:

- the instruction set architecture (ISA) information of the processor;
- the register transfer level (RTL) information of the processor;
- the set of test patterns that must be applied to each of the targeted processor components;
- the self-test coding style that must be applied to each component.

The expected output of such an automatic tool is a set of per-component self-test routines, or a combined, optimized self-test routine for the entire processor which:

- apply the given test patterns to the processor components using instructions;
- guarantee propagation of fault effects to the processor's primary outputs.

For small processor models manual extraction of constraints or manual extraction of processor components information (class, operations, etc) as well as self-test routines development can be quite efficient. It requires the availability of an expert assembly language programmer supplied with the information about the processor instruction set and RT level architecture. This can't be efficiently performed for larger processor models and therefore automation of the software-based self-testing process is necessary for this technology to penetrate the complex SoC market.

We believe that automation of software-based self-testing for processor and SoC architectures will be a research direction with increasing activity within the next few years.

Chapter 6

Case Studies – Experimental Results

In this Chapter we discuss the application of software-based self-testing to several embedded processor designs that are publicly available. As in other cases of engineering research, one of the most difficult problems in software-based self-testing is the availability of reasonably complex and representative benchmark cases to demonstrate the practical value and applicability of the methodology.

Today, the development of embedded processor designs is a profitable business for several fabless companies (see Table 2-2, Table 2-3 and Table 2-4). Therefore, it is very difficult to obtain, for research purposes, a modern processor model with complete functionality. On the other side, it is not necessary, for the demonstration of a methodology, to have an exactly equivalent model of a commercial processor core. It is sufficient to work on a fully functional processor model that realizes the instruction set architecture of a known processor.

The selected processor models described in this Chapter have a wide range of complexities and different instruction set architectures and can be used as benchmarks to demonstrate in a reasonable and persuasive manner the benefits of the software-based self-testing methodology. Their common characteristic is that they are available in a synthesizable, soft core model (VHDL or Verilog). Availability of a synthesizable processor model gives the ability to apply software-based self-testing in different synthesized versions of the processor (optimized for area or delay) and also gives the

Embedded Processor-Based Self-Test
D.Gizopoulos, A.Paschalis, Y.Zorian

ability to calculate the relative sizes of the processor's components. Of course, availability of a synthesizable processor model gives also the ability to perform fault simulation on the synthesized design and obtain fault coverage results that demonstrate the effectiveness of the methodology.

The set of benchmark processor models used in this Chapter can be valuable as a common basis for comparisons in future research in the area of software-based self-testing for processors and processor-based SoC architectures. In any case, the more complex processor models are available, the better for the demonstration of any processor self-testing methodology.

For each of the benchmark processors, after a first brief description, we discuss its implementation/synthesis results providing statistics of the processor's component sizes. Then, we present, for each processor, the fault simulation results from the application of low-cost, software-based self-testing to it. These experimental results include embedded self-test program sizes, self-test execution time, as well as, fault coverage with respect to single stuck-at faults.

We have implemented all benchmark processors using a classic flow of synthesis and simulation from their original VHDL or Verilog source code. A 0.35 um ASIC library[38] has been used for synthesis and gate count is presented in gate equivalents where one gate equivalent is the two-input NAND gate.

We have applied software-based self-testing to all the selected benchmark processors. For each one of the processors, we discuss self-test routines development and we give the self-test program size in words, the response data size in words and execution time in CPU cycles, as well as, the fault coverage obtained for the entire processor and its individual components (those that have been targeted for test development or not).

6.1 Parwan processor core

Parwan [116] is a very simple 8-bit accumulator-based processor that has been developed for educational purposes and is briefly mentioned in this section only because it has been used as a demonstration vehicle in the software-based self-testing methodologies presented in [27], [28], [29], [94], [95], [96], [104], [105]. Parwan is an 8-bit CPU with a 12-bit address bus able to access a 4K memory. The Parwan instruction set includes common instructions like load from and store to memory, arithmetic and logical operations, jump and branch instructions and is therefore able to implement several real algorithms. Parwan also supports direct and indirect addressing

[38] A second 0.50 um ASIC library has been also used for comparisons between two different libraries. This comparison has been performed for the Plasma/MIPS processor benchmark.

modes for memory operands. Considering that each addressing mode leads to different instructions, Parwan's instruction set consists of 24 different instructions.

The Parwan processor model is available in VHDL synthesizable format and its architecture includes the components shown in Table 6-1. The classification of each component to the classes described in the previous Chapter is also shown in Table 6-1.

Component Name	Component Class
Arithmetic Logic Unit (ALU)	Functional computational
Shifter Unit (SHU)	Functional computational
Accumulator (ACC)	Functional storage
Program Counter (PC)	Control
Status Register (SR)	Control
Memory Address Register (MAR)	Control
Instruction Register (IR)	Control
Control Unit (CRTL)	Control

Table 6-1: Parwan processor components.

Out of the eight processor components only the Arithmetic Logic Unit and the Shifter are combinational circuits and also the only functional computational components. It should also be noted that the only processor data register is the accumulator. This single data processor register is fully accessible in terms of controllability and observability by processor instructions.

We have synthesized Parwan from its VHDL source description and the resulting circuit consists of 1,300 gates including 53 flip-flops.

6.1.1 Software-based self-testing of Parwan

The Parwan processor components that have been targeted for self-test program development are the ALU, the Shifter and the Status Register. Self-test routines development has been done in three phases: Phase A for the ALU, Phase B for the Shifter and Phase C for the Status Register. We have selected this sequence because the third functional unit of the processor, the Accumulator, is already sufficiently tested after Phase A.

Table 6-2 shows the statistics of the self-test code for the three consecutive phases A, B and C.

	Phase A *target ALU*	**Phase B** *target Shifter*	**Phase C** *target Status Register*
Number of Instructions	311	440	463
Self-Test Program Size (bytes)	631	881	923
Response Data Size (bytes)	72	122	124
Execution Time (cycles)	9,154	16,545	16,667

Table 6-2: Self-test program statistics for Parwan.

Table 6-3 shows the fault coverage for single stuck-at faults for each of the Parwan processor components after each of the three phases.

Component Name	**Fault Coverage after Phase A**	**Fault Coverage after Phase B**	**Fault Coverage after Phase C**
Arithmetic Logic Unit (ALU)	98.31%	98.48%	98.48%
Shifter Unit (SHU)	75.56%	93.82%	93.82%
Accumulator (ACC)	98.67%	98.67%	98.67%
Program Counter (PC)	87.05%	88.10%	88.10%
Status Register (SR)	92.13%	92.13%	92.13%
Memory Address Register (MAR)	97.22%	97.22%	97.22%
Instruction Register (IR)	98.26%	98.26%	98.26%
Control Unit (CRTL)	82.93%	85.52%	87.68%
Total CPU	88.70%	91.10%	91.34%

Table 6-3: Fault simulation results for Parwan processor.

Experimental results on Parwan for single stuck-at faults were given in [28]. Single stuck-at fault coverage of 91.42% is reported in [28] which was obtained by a self-test program of 1,129 bytes that needs 137,649 clock cycles for execution (compare with the 923 bytes and the only 16,667 clock cycles reported in Table 6-3).

6.2 Plasma/MIPS processor core

Plasma [128] is a publicly available RISC processor model that implements the MIPS I™ instruction set architecture. Plasma supports interrupts and all MIPS I™ user mode instructions except unaligned load and store operations (which are patented) and exceptions.

The synthesizable CPU core is implemented in VHDL with a 3-stage pipeline structure. The Plasma processor architecture consists of the components shown in Table 6-4. We can also see in the list below the characterization of each of the components into the classes described in the previous Chapter.

Component Name	Component Class
Register File	Functional storage
Multiplier	Functional computational
Divider	Functional computational
Arithmetic-Logic Unit	Functional computational
Shifter	Functional computational
Memory Control	Control
Program Counter Logic	Control
Control Logic	Control
Bus Multiplexer	Functional interconnect
Pipeline	Hidden

Table 6-4: Plasma processor components.

The internal architecture of each of the processor components can be of many different types. In particular, the components that perform arithmetic operations such as the ALUs, adders, subtracters, multipliers, dividers, incrementers, etc can be designed in many different ways. For the case of the Plasma/MIPS processor benchmark, we have implemented two different architectures of the multiplier component, a serial one and a parallel one, each of which leads to different gate count for the processor. As we see in the experimental results, the low-cost, software-based self-testing methodology is able to obtain very high fault coverage in both cases.

For comparison purposes, we have synthesized the Plasma/MIPS processor core into three different designs which we call Design I, II and III, respectively. Design I contains a serial multiplier, Design II contains a parallel multiplier and synthesis has been performed for area optimization, while Design III also contains a parallel multiplier but it is a delay optimized design. Table 6-5, Table 6-6 and Table 6-7 show the synthesis results for these three implementations of Plasma. Design I operates in a 66.0 MHz frequency, Design II in a 57.0 MHz frequency and Design III in a 74.0 MHz frequency.

The total gate count of the synthesized processor is slightly larger than the sum of gate counts of the individual components due to the existence of glue logic among the processor components, at the top level of hierarchy, which can't be identified as a separate processor component.

Component Name	Gate Count
Register File	9,906
Multiplier/Divider [39]	3,044
Arithmetic-Logic Unit	491
Shifter	682
Memory Control	1,112
Program Counter Logic	444
Control Logic	223
Bus Multiplexer	453
Pipeline	885
Total CPU	17,459

Table 6-5: Plasma processor synthesis for Design I.

Component Name	Gate Count
Register File	9,905
Multiplier/Divider [40]	11,601
Arithmetic-Logic Unit	491
Shifter	682
Memory Control	1,119
Program Counter Logic	444
Control Logic	230
Bus Multiplexer	453
Pipeline	885
Total CPU	26,080

Table 6-6: Plasma processor synthesis for Design II.

[39] In Design I, multiplier/divider are together implemented in a single component and in serial architecture.

[40] In Designs II and III, multiplier/divider are again a single component but the multiplier has a parallel implementation while the divider keeps its serial structure. This is a very natural choice because the division operation is a very rare one compared to multiplication, and therefore only the multiplication operation deserves a parallel, more efficient implementation.

Component Name	Gate Count
Register File	11,905
Multiplier/Divider	13,358
Arithmetic-Logic Unit	900
Shifter	834
Memory Control	1,163
Program Counter Logic	493
Control Logic	361
Bus Multiplexer	623
Pipeline	961
Total CPU	30,896

Table 6-7: Plasma processor synthesis for Design III.

We note that in the three implementations of the Plasma processor the total gate counts of its functional components represent an 83.48% (Design I), 88.69% (Design II) and 89.39% (Design III) of the entire processor's area, respectively. Therefore, in this processor benchmark, our claim that the functional components of the processor are the largest and therefore those that must be initially targeted by low-cost, software-based self-testing, is easily justified. We will see that in the other processor benchmarks the same argument is still valid to a smaller or larger extend.

6.2.1 Software-based self-testing of Plasma/MIPS

We have applied low-cost, software-based self-testing to Plasma/MIPS processor benchmark. For Design I we have developed a self test program targeting the Register File, the Arithmetic Logic Unit, the Multiplier/Divider, the Shifter and the Memory Control components. The self-test program consists of 965 words (32-bits) and is executed in a total of 3,552 clock cycles. The fault coverage obtained for each of the processor components is shown in Table 6-8.

Component Name	Fault Coverage for Design I
Register File	97.7%
Multiplier/Divider	87.5%
Arithmetic-Logic Unit	96.6%
Shifter	98.4%
Memory Control	88.3%
Program Counter Logic	53.1%
Control Logic	78.9%
Bus Multiplexer	65.7%
Pipeline	91.9%
Total CPU	92.2%

Table 6-8: Fault simulation results for the Plasma processor Design I.

The next two implementations of the Plasma processor (Design II and Design III) contain a parallel multiplier instead of a serial one. The inclusion of a parallel multiplier in the processor increases its size (see Table 6-6 and Table 6-7) by more than 8,000 gates (32 bit array multiplier) but it speeds up the multiplication operation which is now executed in a single cycle instead of 32 of the serial implementation. Design II is optimized for area and Design III is optimized for speed, so the area of Design III is larger by about 4,000 gates for the entire processor.

For these two implementations of Plasma, we have developed the same self-test routines for the following components: Register File, Parallel Multiplier, Serial Divider, Arithmetic-Logic Unit, Shifter, Memory Control, and Control Logic. The self-test routine sizes and execution times for each component and the overall processor are shown in Table 6-9.

Targeted Component	Self-Test Routine Size (words)	Execution Time (cycles)
Register File	319	582
Parallel Multiplier	28	3,122
Serial Divider	41	1,154
Arithmetic-Logic Unit	79	275
Shifter	210	340
Memory Control	76	160
Control Logic	100	164
Total CPU	853[41]	5,797

Table 6-9: Self-test routine statistics for Designs II and III of Plasma.

41 In Design I with the serial multiplier the total program was 965 words and total cycles were 3,552 because the serial multiplier needs a larger test program that is executed for more clock cycles than the one for the parallel multiplier.

We notice in the self-test routine statistics of Table 6-9 that there exist components with very small self-test routines, such as the multiplier and the divider, which routines take a very large percentage of the overall test execution time. This is because these routines consist of small, compact loops that are executed for a large number of times and apply a large number of test patterns to the component they test. On the other side, there are components like the Shifter with a relatively large self-test code which is executed very fast because it is a code that does not contain loops but rather applies a small set of test patterns using immediate addressing mode instructions for all shifter operations. The self-test routine for the Arithmetic-Logic Unit consists of segments for every ALU operation that combine small compact loops and immediate addressing mode instructions. The self-test routine of the Memory Control consists of load instructions with specific data previously stored in data memory, and store instructions that generates the final responses in data memory as well. Finally, the self-test routine of the Control Logic component is based on an exhaustive application of all the processor's instruction opcodes not already applied in the routines of the previous components. This functional testing approach at the end of the methodology is very natural to be applied to the control unit of the processor.

At this point we remark that a similar self-test routine development strategy has been adopted for the remaining benchmark processors, for their components that are similar with the components of Plasma/MIPS.

The fault simulation results for the two designs of Plasma/MIPS, Design II and III are shown in Table 6-10.

Component Name	Fault Coverage for Design II	Fault Coverage for Design III
Register File	97.8%	97.8%
Multiplier/Divider	96.3%	95.2%
Arithmetic-Logic Unit	96.8%	95.8%
Shifter	98.4%	95.3%
Memory Control	87.9%	90.3%
Program Counter Logic	54.9%	55.9%
Control Logic	89.3%	85.3%
Bus Multiplexer	71.8%	71.3%
Pipeline	98.4%	96.0%
Total CPU	95.3%	94.5%

Table 6-10: Fault simulation results for Designs II and III of Plasma.

We see that in the two implementations with the parallel multiplier the overall fault coverage is higher than in the case of Design I which contains a serial implementation of the multiplier. This fault coverage increase is simply due to the fact that a large component like the parallel multiplier with very high testability has been inserted in the processor. The same design change could be done for the division operation and another large component, the parallel divider, could be added. This would lead to further increase of the processor's fault coverage. We didn't implement this design change because the division operation is not as common as the multiplication and therefore the cost of adding a large parallel component that is infrequently used is not justified in low-cost applications. Of course in a special implementation of the processor in an application with many divisions, the inclusion of a parallel divider will lead to an increased performance of the processor, as well as, an increased fault coverage obtained by software-based self-testing. In such a case, where a parallel divider is also implemented, fault coverage can be as high as 98% for single stuck-at faults.

The overall processor fault coverage is, in both designs, very high (more than 94.5%) while the particular fault coverage levels of each component may slightly differ because of the different synthesis optimization parameters that lead to a different gate-level implementation of the component.

We also note that the pipeline logic is tested as a side-effect of testing the remaining processor components achieving very high fault coverage. This fact is due to the simple pipeline structure that Plasma realizes.

6.2.1.1 Application to different synthesis libraries

We have performed another experiment with the Plasma processor benchmark in order to evaluate the effect of synthesizing the processor with a different implementation library. A 0.50 um library has been used and synthesis has been done with area optimization. We call this synthesized design as Design IV. Design IV contains a parallel multiplier, has a frequency of 42.0 MHz and its synthesis results in gate counts are given in Table 6-11.

Component Name	Gate Count
Register File	11,558
Multiplier/Divider	11,654
Arithmetic-Logic Unit	558
Shifter	636
Memory Control	1,120
Program Counter Logic	449
Control Logic	244
Bus Multiplexer	431
Pipeline	876
Total CPU	27,824

Table 6-11: Plasma processor synthesis for Design IV.

As Design IV is an area optimized one, it is directly comparable with Design II. We applied to Design IV exactly the same self-test routines (Table 6-9) that were applied to Design II (and also to Design III) and the comparison of fault coverages for all components in Designs II and IV are summarized in Table 6-12.

Component Name	Fault Coverage for Design II	Fault Coverage for Design IV
Register File	97.8%	97.8%
Multiplier/Divider	96.3%	96.1%
Arithmetic-Logic Unit	96.8%	97.5%
Shifter	98.4%	99.9%
Memory Control	87.9%	88.5%
Program Counter Logic	54.9%	54.9%
Control Logic	89.3%	88.3%
Bus Multiplexer	71.8%	72.0%
Pipeline	98.4%	96.9%
Total CPU	95.3%	95.3%

Table 6-12: Comparisons between Designs II and IV of Plasma.

We note that the total fault coverage for the processor is *exactly* the same 95.3% of all single stuck-at faults. For each of the components of the processor, fault coverage results may slightly differ to a maximum of 1.5% of the component's faults (shifter, pipeline). These small differences are due to the different cells that each implementation library contains. The same self-test program reaches slightly different structural fault coverage in each of the processor components.

Some useful conclusions can be drawn from this first application of software-based self-testing to the reasonable size processor model:

- Self-test code size is very small, less than 1,000 words.

- Self-test routines execution time is also very small, less than 6,000 clock cycles.
- The fault coverage for the largest functional storage component, the Register File, is always very high.
- The fault coverage for all the functional computational components that have been targeted is very high.
- The pipeline logic is sufficiently tested although it has not been targeted for self-test code development. This fact is due to the simple pipeline structure of Plasma.

6.3 Meister/MIPS reconfigurable processor core

Our next selected CPU benchmark that belongs to the class of reconfigurable CPUs which have gained importance today because they offer the ability to tailor an instruction set and corresponding functional units of a processor for a particular application. Based on the Meister Application Specific Instruction set Processor (ASIP) environment [114], we have designed a MIPS R3000 (MIPS I™ ISA) compatible processor with a 5-stage pipeline structure. A subset of 52 instructions of the MIPS R3000 instruction set [85] was implemented while co-processor and interrupt instructions were not implemented in this experiment. It must be mentioned that in the current educational release version of the ASIP design environment of [114] data hazard detection and forwarding are not completely implemented. Therefore, although the pipeline structure exists (pipeline registers, multiplexers, etc) its complete functionality is not used. An RTL VHDL model was generated by the tool for this MIPS compatible processor.

The Meister/MIPS processor consists of the identifiable components shown in Table 6-13. We can also see in the list below the characterization of each of the components into the classes described in the previous Chapter.

Component Name	Component Class
Register File	Functional storage
Parallel Multiplier/Serial Divider	Functional computational
ALU	Functional computational
Shifter	Functional computational
Hi-Lo Registers	Functional storage
Controller	Control
Data Memory Controller	Control
Program Counter Logic	Control
Instruction register	Hidden
Pipeline registers	Hidden

Table 6-13: Meister/MIPS processor components.

The usefulness of experimenting with Meister/MIPS is that we can compare the results of software-based self-testing in this architecture which is another implementation of about the same instruction set of the previous model, Plasma/MIPS. Meister/MIPS configurable processor is another implementation of the classical, popular instruction set architecture of MIPS that the Plasma/MIPS benchmark also implements. We performed our experiments with the Meister/MIPS too, to verify the effectiveness of the approach to another implementation of the same instruction set architecture where the components of the processor are internally implemented in a different way than Plasma/MIPS. Table 6-14 shows the synthesis results of the Meister/MIPS processor. The clock frequency of the Meister/MIPS implementation is 44.0 MHz.

Component Name	Gates Count
Register File	11,414
Parallel Multiplier/Serial Divider	12,564
ALU	658
Shifter	633
Hi-Lo Registers	536
Controller	2,352
Data Memory Controller	1,086
Program Counter Logic	644
Instruction register	275
Pipeline registers	5,693
Total CPU	37,402

Table 6-14: Meister/MIPS processor synthesis.

As in the case of Designs II and III of Plasma, the component for multiplication and division is a single component that contains a parallel multiplier and a serial divider. A slight difference between Plasma/MIPS and Meister/MIPS is that the Hi and Lo registers that hold the results of multiplication and division in the MIPS architecture are implemented as a separate component in Meister/MIPS, while in Plasma/MIPS they are part of the multiplier/divider component.

We can see that in Meister the functional components of all types occupy a total of 68.99% of the total processor's area. This is still a very large percentage which is smaller than in the case of Plasma/MIPS because Meister/MIPS implements a more complex pipeline structure where pipeline registers and other logic occupy a significantly larger area of the processor than in the case of Plasma/MIPS.

6.3.1 Software-based self-testing of Meister/MIPS

We have developed self-test routines for the list of the components shown in Table 6-15 which together compose a self-test program of 1,728 words executed in 10,061 clock cycles for the Meister/MIPS processor. We can also see the self-test routine size and execution time for each component.

Targeted Component	Self-Test Routine Size (words)	Execution Time (cycles)
Register File	720	859
Parallel Multiplier	68	5,855
Serial Divider	65	1,396
Arithmetic-Logic Unit	192	1,188
Shifter	378	437
Hi-Lo Registers	30	35
Control Logic	275	291
Total CPU	1,728	10,061

Table 6-15: Self-test routines statistics for Meister/MIPS processor.

The fault coverage obtained by the above self-test routines for Meister/MIPS after evaluation of 600 responses in data memory is given in Table 6-16.

Component Name	Fault Coverage
Register File	99.8%
Parallel Multiplier/Serial Divider	95.2%
ALU	98.4%
Shifter	99.8%
Hi-Lo Registers	100.0%
Controller	79.2%
Data Memory Controller	58.2%
Program Counter Logic	58.5%
Instruction register	97.4%
Pipeline registers	91.0%
Total CPU	92.6%

Table 6-16: Fault simulation results for Meister/MIPS processor.

We note that the Meister/MIPS processor benchmark implements almost the same MIPS I™ instruction set architecture as the Plasma model, but with a more complex control and pipeline. This is due to the fact that the VHDL RTL design generated by the ASIP design environment does not support data hazard detection and forwarding. In the Meister/MIPS processor, careful assembly instruction scheduling is required along with insertion of nop (no operation) instructions wherever it is necessary. The result is an increased test program size (almost double) and test execution time (almost double) and smaller fault coverage (92.6% compared to more than 94.5% of Plasma). Otherwise (if the pipeline logic was completely implemented) test program statistics would have been very similar to the case of Plasma/MIPS and fault coverage would have been much higher.

6.4 Jam processor core

Jam CPU [78] has been implemented in Chalmers University of Technology. It follows a 32-bit RISC CPU architecture called Concert'02. Jam CPU has a five-stage pipeline structure which is very similar to the pipeline structure of the DLX architecture [63]. Jam CPU is implemented in VHDL synthesizable format and its five-stage pipeline structure includes the following stages: Instruction Fetch, Instruction Decode, Execute, Memory and Write Back. Jam has multi-cycle operations, pipeline hazard checking and pipeline forwarding.

Jam processor consists of the identifiable components shown in Table 6-17. We can also see in the list below the characterization of each of the components into the classes described in the previous Chapter.

Component Name	Component Class
Register File (REGS)	Functional storage
Integer Unit (IU) – *includes the ALU*	Functional computational
Immediate Extender (IMM_EXT)	Functional computational
Memory Access Unit (MAU 1) for Instruction Memory	Control
Memory Access Unit (MAU 2) for Data Memory	Control
Control Logic (CONTROL)	Control
Pipeline Registers	Hidden

Table 6-17: Jam processor components.

The usefulness of experimenting with Jam CPU is that is a step forward to the study of software-based self-testing in RISC architectures. It is a different architecture than the one implemented by Plasma/MIPS and Meister/MIPS and it contains a fully implemented, five-stage pipeline structure with hazard detection and forwarding.

The pipeline structure of the Jam processor core is implemented at the top level of the VHDL design. It is therefore difficult to identify components that implement the pipeline mechanism other than the pipeline registers. The rest of the pipeline logic (control logic for hazard detection and multiplexers for forwarding) is not accounted as separate components. The synthesis results of the Jam processor from its original VHDL source code are given in Table 6-18. The clock frequency of the Jam processor implementation is 41.8 MHz.

Component Name	Gate Count
Register File (REGS)	22,917
Integer Unit (IU)	5,698
Immediate Extender (IMM_EXT)	269
Memory Access Unit (MAU 1) for Instruction Memory	576
Memory Access Unit (MAU 2) for Data Memory	576
Control Logic (CONTROL)	388
Pipeline Registers	3,771
Total CPU	43,208

Table 6-18: Jam processor synthesis.

The register file consists of 32 registers of 32 bits and is the largest component of the processor that dominates the processor area (more than 50%). The Integer Unit in the Jam processor core implements the following integer operations: multiplication, addition, subtraction, bitwise OR, bitwise AND, bitwise XOR, shift. All operations apart from multiplication take one cycle, while multiplication takes 33 cycles (multiplication in the Jam processor integer unit is implemented in a serial fashion, like in the case of the Plasma Design I). The Integer Unit contains control logic, a shift register used for multiplication and an ALU that handles addition, subtraction and the logical operations. The Immediate Extender participates in the immediate addressing mode instructions. The Memory Access Units 1 and 2 are used for the communication of the processor with the instruction memory and data memory, respectively.

The remaining gates in the Jam processor not given in Table 6-18 are part of the top-level entity and include not only glue logic, but pipeline multiplexers and control used for hazard detection and forwarding. They represent a total of about 9,000 gates.

We can see that the functional components in the case of the Jam processor occupy a total of 66.85% of the processor's area. This fact is again (as in Meister/MIPS processor) a result of the complex pipeline structure of the processor and all the logic gates needed to implemented it (only the pipeline registers, which are the identifiable part of the processor's pipeline architecture occupy more than 8% of the processor).

6.4.1 Software-based self-testing of Jam

The Jam processor benchmark gives us the ability to evaluate software-based self-testing in a more complex RISC processor model with a fully implemented pipeline architecture which realizes hazard detection and forwarding.

We have developed self-test routines for the list of components shown in Table 6-19. In this Table we can also see the self-test routine size and execution time for each component. These routines together compose a self-test program of 897 words executed in 4,787 clock cycles for the Jam processor.

Component Name	Self-Test Routine Size (words)	Execution Time (cycles)
Register File (REGS)	478	550
Integer Unit (IU)	147	3,920
Immediate Extender (IMM_EXT)	32	38
Memory Access Unit (MAU 2)	120	135
Control Logic (CONTROL)	120	144
Total CPU	897	4,787

Table 6-19: Self-test routine statistics for Jam processor.

We have not developed any special self-test routines for the pipeline logic since pipeline forwarding is already activated many times during the execution of the self-test program and the pipeline logic, multiplexers and registers are sufficiently tested as a side-effect of testing the remaining components. The fault simulation results for the Jam processor after evaluation of 454 responses in data memory are shown in Table 6-20. The achieved overall processor fault coverage is very high, 94% with respect to single stuck-at faults.

Component Name	Fault Coverage
Register File (REGS)	98.1%
Integer Unit (IU)	98.9%
Immediate Extender (IMM_EXT)	98.5%
Memory Access Unit (MAU 1) for Instruction Memory	69.4%
Memory Access Unit (MAU 2) for Data Memory	81.7%
Control Logic (CONTROL)	81.2%
Pipeline Registers	89.7%
Total CPU	94.0%

Table 6-20: Fault simulation results for Jam processor.

6.5 oc8051 microcontroller core

The 8051 microcontroller is a member of MCS-51 family, originally designed in the 1980's by Intel. The 8051 has gained great popularity since its introduction and, it is estimated that it is used in a large percentage of all embedded system products today.

We have selected as a benchmark the oc8051 model of the 8-bit 8051 microcontroller [121]. This implementation of 8051 has a two stage pipeline structure (major difference with other 8051 implementations). At the first pipeline stage instruction fetching and decoding takes place while during second pipeline stage the instructions are executed and the results are stored in memory. The oc8051 model is implemented in Verilog HDL.

The oc8051 processor consists of the identifiable components shown in Table 6-21. We can also see in the list below the characterization of each of the components into the classes described in the previous Chapter.

Component Name	Component Class
Arithmetic Logic Unit (ALU)	Functional computational
ALU Source Select (ASS)	Functional interconnect
Decoder (DEC)	Control
Special Function Registers (SFR)	Functional storage
Indirect Address (INDI_ADDR)	Functional storage
Memory Interface (MEM)	Control
Pipeline Registers	Hidden

Table 6-21: oc8051 processor components.

The usefulness of experimenting with the oc8051 model is that it is a very popular architecture used extensively as an embedded processor core in SoCs. It follows a classical Intel architecture and is a substantially different design from the RISC processors described in the previous sections. The synthesis results of oc8051 from its original Verilog source code are presented in following Table 6-22. The clock frequency of the oc8051 processor implementation is 83.1 MHz.

Component Name	Gate Count
Arithmetic Logic Unit (ALU)	1,147
ALU Source Select (ASS)	269
Decoder (DEC)	970
Special Function Registers (SFR)	4,507
Indirect Address (INDI_ADDR)	635
Memory Interface (MEM)	2,703
Total CPU	10,305

Table 6-22: oc8051 processor synthesis.

The Arithmetic Logic Unit in the oc8051 processor core implements the following integer operations: addition, subtraction, bitwise OR, bitwise AND, bitwise XOR, shift and multiplication.

The ALU Source Select component selects the ALU input sources. The Special Function Registers component contains 18 registers of one or two bytes each (accumulator, B register, Program Status Word, Stack Pointer, etc). The Indirect Address component implements the indirect addressing mode instructions. The Memory Interface component is used for the communication of the processor with the memory, while the Decoder component is the control logic of the processor. The remaining logic gates of oc8051 not included in Table 6-22 are again part of the top level of the hierarchy, constitute the logic for the implementation of the 2-stage pipeline structure of oc8051 and other surrounding logic to the above components.

We see that in the case of the oc8051 processor a total of 63.64% of its area is occupied by the functional components. Targeting these components for self-test routines development and also considering the simple pipeline structure of the processor, we have good chances to obtain high fault coverage with small self-test code.

6.5.1 Software-based self-testing of oc8051

In the case of the oc8051 microcontroller benchmark we have developed self-test routines for the list of components shown in Table 6-23. In Table 6-23 we can also see the self-test routine size and execution time for each component and the overall processor. These routines together compose a self-test program of 3,760 bytes executed in 5,411 clock cycles for the oc8051 microcontroller.

Component Name	Self-Test Code Size (bytes)	Self-Test Code Time (cycles)
Arithmetic Logic Unit (ALU)	1,452	2,964
ALU Source Select (ASS)	512	541
Special Function Registers (SFR)	548	598
Indirect Address (INDI_ADDR)	560	614
Decoder (DEC)	360	324
Memory Interface (MEM)	328	370
Total CPU	3,760	5,411

Table 6-23: Self-test routine statistics for oc8051 processor.

The fault simulation results after evaluation of 416 test responses in memory are shown in Table 6-24.

Component Name	Fault Coverage
Arithmetic Logic Unit (ALU)	98.4%
ALU Source Select (ASS)	97.1%
Decoder (DEC)	81.5%
Special Function Registers (SFR)	96.2%
Indirect Address (INDI_ADDR)	90.9%
Memory Interface (MEM)	89.9%
Total CPU	93.1%

Table 6-24: Fault simulation results for oc8051 processor.

We see that for oc8051 as well high fault coverage of 93.1% with respect to single stuck-at faults is obtained with a relatively small and fast program.

6.6 RISC-MCU microcontroller core

RISC-MCU [136] is a RISC microcontrol unit designed after the AVR 8-bit RISC microcontroller from Atmel. It is a synthesizable processor core model implemented in VHDL. RISC-MCU has the same instruction set with the Atmel AT90S1200 processor (89 instructions).

RISC-MCU microcontroller consists of the identifiable components shown in Table 6-25. We can also see in the list below the characterization of each of the components into the classes described in the previous Chapter.

Component Name	Component Class
Shift Register (SR)	Functional computational
General Purpose Registers (GPR)	Functional storage
Arithmetic Logic Unit (ALU)	Functional computational
Control Unit (CTRL)	Control
Program Counter (PC)	Control
IO Port (IOP)	Control
Timer	Control
Pipeline	Hidden

Table 6-25: RISC-MCU processor components.

The usefulness of experimenting with the AVR model provided by RISC-MCU is to compare it with the previous RISC processor models described in the previous sections. RISC-MCU is a smaller word (8-bit) RISC architecture as opposed to the other larger word (32-bit) architectures that we studied so far. The results obtained from its synthesis from the original VHDL source files are given in the following Table 6-26. The clock frequency of the RISC-MCU processor implementation is 249.9 MHz (much faster than the 32-bit RISC processors).

Component Name	Gate Count
Shift Register (SR)	127
General Purpose Registers (GPR)	1,513
Arithmetic Logic Unit (ALU)	406
Control Unit (CTRL)	777
Program Counter (PC)	645
IO Port (IOP)	185
Timer	294
Pipeline	650
Total CPU	4,693

Table 6-26: RISC-MCU processor synthesis.

This smaller RISC processor benchmark has almost the same functional components of larger 32-bit RISC processors but for a smaller word length. We can see that the functional components of all types in RISC-MCU occupy a total of 43.60% of the entire processor area. This percentage is much smaller than in the case of larger 32-bit RISC processors described in the previous sections. This is due to the fact that although the control logic components remain of the same order of magnitude for a small word length (8 bits) or a large word length (32 bits), the datapath components that perform the operations (computational functional components) and store the data (storage functional components) are significantly larger in the 32-bits word length. For example, the General Purpose Registers (GPR) component of RISC-MCU occupies only 1,513 gates, while the corresponding register files of Plasma, Meister and Jam occupy more than 9,000, 11,000 or even 22,000 gates (Jam).

6.6.1 Software-based self-testing of RISC-MCU

For the RISC-MCU benchmark we have developed self-test routines for the list of components shown in Table 6-27. In Table 6-27 we can also see the self-test routine size and execution time for each component and the overall processor. These routines together compose a self-test program of 1,258 words executed in 2,446 clock cycles for the RISC-MCU microcontroller.

Component Name	Self-test Code Size (words)	Self-test Code Time (cycles)
Shift Register (SR)	360	410
General Purpose Registers (GPR)	510	674
Arithmetic Logic Unit (ALU)	128	1,032
Control Unit (CTRL)	200	240
IO Port (IOP)	60	90
Total CPU	1,258	2,446

Table 6-27: Self-test routine statistics for RISC-MCU processor.

The fault simulation results for the RISC-MCU microcontroller are shown in Table 6-28.

Component Name	Fault Coverage
Shift Register (SR)	99.2%
General Purpose Registers (GPR)	99.0%
Arithmetic Logic Unit (ALU)	98.4%
Control Unit (CTRL)	91.1%
Program Counter (PC)	72.5%
IO Port (IOP)	91.4%
Timer	74.0%
Pipeline	92.4%
Total CPU	91.2%

Table 6-28: Fault simulation results for RISC-MCU processor.

We see that high fault coverage of 91.2% with respect to single stuck-at faults is obtained with a relatively small and fast self-test program. The fault coverage is less than in the case of larger 32-bit RISC processors due to the smaller scale of the functional components.

6.7 oc54x DSP Core

The oc54x DSP core is a synthesizable 16/32, dual-16 bit DSP core which is available in synthesizable Verilog format [120]. Oc54x is an implementation of a popular family of DSPs designed by Texas Instruments and is software compliant with the original TI C54x DSP.

The oc54x DSP processor consists of the identifiable components shown in Table 6-29. We can also see in the list below the characterization of each of the components into the classes described in the previous Chapter.

Component Name	Component Class
Accumulator (ACC)	Functional computational
Arithmetic Logic Unit (ALU)	Functional computational
Barrel Shifter (BSFT)	Functional computational
Compare/Select and Store Unit (CSSU)	Functional interconnect
Exponent Decoder (EXP)	Functional computational
Multiply/Accumulate Unit (MAC)	Functional computational and storage
Temporary Register (TREG)	Functional storage
Control	Control

Table 6-29: oc54x processor components.

The usefulness of experimenting with the oc54x DSP model is that it is an implementation of a popular DSP architecture. DSPs are very widely used in SoC architectures either alone or in conjunction with other general purpose processor cores. The oc54x DSP core, like all DSPs are excellent candidates for the application of low-cost, software-based self-testing, because they contain large word length functional components (mostly computational and storage ones). The synthesis results for oc54x from its original Verilog source code are given in Table 6-30. The clock frequency of the oc54x DSP implementation is 48.8 MHz.

Component Name	Gates Count
Accumulator (ACC)	687
Arithmetic Logic Unit (ALU)	3,145
Barrel Shifter (BSFT)	1,682
Compare/Select and Store Unit (CSSU)	444
Exponent Decoder (EXP)	799
Multiply/Accumulate Unit (MAC)	3,819
Temporary Register (TREG)	130
Control	905
Total CPU	11,611

Table 6-30: oc54x DSP synthesis.

We can see that the synthesis statistics of the oc54x DSP make it a very suitable architecture for the application of software-based self-testing because of the existence of many and large functional components. A total of 92.21% of the entire DSP area is occupied by the functional units of all subtypes. All of them are well accessible units that can be easily targeted with software-based self-testing routines.

6.7.1 Software-based self-testing of oc54x

Our DSP benchmark oc54x has very high fault coverage because it consists of many and large functional components and a small control logic. We have developed self-test routines for the following functional components of oc54x as shown in Table 6-31. In this list we can also see the self-test routine size and execution time for each component and the overall DSP. These routines together compose a self-test program of 1,558 words executed in 7,742 clock cycles for the oc54x DSP benchmark.

Component Name	Self-test Routine Size (words)	Execution Time (cycles)
Accumulator (ACC)	64	78
Arithmetic Logic Unit (ALU)	102	956
Barrel Shifter (BSFT)	700	778
Compare/Select and Store Unit (CSSU)	180	264
Exponent Decoder (EXP)	256	340
Multiply/Accumulate Unit (MAC)	24	5,020
Temporary Register (TREG)	112	152
Control	120	154
Total CPU	1,558	7,742

Table 6-31: Self-test routines statistics for oc54x DSP.

The targeted functional components of the oc54x DSP are very classical components in a DSP datapath and the generation of self-test routines for them is similar to other components like these for the previous processors' benchmarks. The fault simulation results after evaluation of 572 test responses in data memory are shown in Table 6-32.

We see that very high fault coverage of 96.0% with respect to single stuck-at faults is obtained with a relatively small and fast self-test program. The very high fault coverage is achieved because the vast majority of processor area is occupied by its functional components.

Component Name	Fault Coverage
Accumulator (ACC)	99.2%
Arithmetic Logic Unit (ALU)	98.0%
Barrel Shifter (BSFT)	99.4%
Compare/Select and Store Unit (CSSU)	89.0%
Exponent Decoder (EXP)	91.1%
Multiply/Accumulate Unit (MAC)	98.9%
Temporary Register (TREG)	98.0%
Control	84.0%
Total CPU	96.0%

Table 6-32: Fault simulation results for oc54x DSP.

6.8 Compaction of test responses

For each of the processor benchmarks we have also performed compaction of the test responses which are stored in data memory, so that a single final signature is collected. For this reason, an additional compaction routine of few instructions has been developed (the size of the routine is from 10 to 20 words depending on the benchmark). When compaction is applied, the self-test program is slightly modified by changing the store instructions that store the components' test responses in the data memory with a call instruction that transfers the control to the compaction routine. The application of the compaction scheme increases significantly the test execution time of the modified test program. Table 6-33 shows the execution times for the benchmark processors with and without compaction routines executed. The times for the execution without compaction are collected from the Tables of the previous sections.

Benchmark Processor	Execution time without compaction (clock cycles)	Execution time with compaction (clock cycles)
Plasma (Designs II, III, IV)	5,797	9,874
Meister	10,061	23,865
Jam	4,787	8,876
oc8051	5,411	9,513
RISC-MCU	2,446	5,122
oc54x	7,742	12,893

Table 6-33: Execution times of self-test routines.

The application of compaction slightly affects the size of the self-test program (by the 10 to 20 instructions we mentioned), but it seriously increases the test application time. The application of compaction during manufacturing testing depends on the frequency of the external tester. If the time for uploading the responses is less than the time required for compaction then compaction is not performed. During on-line testing, where no external tester exists in field, the application of compaction is necessary.

6.9 Summary of Benchmarks

The following Table 6-34 summarizes the processor cores discussed in this Chapter. The last column of Table 6-34 gives a statement for the usefulness of experimenting with each of these processor benchmarks.

Benchmark Processor	Description	HDL	Usefulness
Parwan	8-bit, accumulator-based processor	VHDL	*Proof of concept. Has been used in related literature.*
Plasma/ MIPS	32-bit RISC processor	VHDL	*Classic 32-bit RISC architecture. Used for different synthesis objectives.*
Meister/ MIPS	32-bit RISC processor	VHDL	*Same classic 32-bit RISC architecture implemented by an ASIP environment.*
Jam	32-bit RISC processor	VHDL	*Another 32-bit RISC architecture. Larger design because of more complex pipeline structure.*
oc8051	8-bit microcontroller	Verilog	*Classic, 8-bit accumulator-based architecture. Very popular embedded core.*
RISC-MCU	8-bit RISC microcontroller	VHDL	*A modern, small and fast 8-bit RISC architecture for comparison with 32-bit RISC ones.*
oc54x	16/32, dual 16-bit DSP	Verilog	*A DSP architecture from a successful model.*

Table 6-34: Summary of benchmark processor cores.

We remark that all the benchmarks we used in this Chapter are publicly available and represent some very good efforts to implement some common, classic and popular instruction set architectures. The success of the application of software-based self-testing in these benchmarks gives a strong evidence for the practicality and usefulness of the methodology, but this success does not mean that the approach can be applied straightforwardly to any commercial embedded processor, microprocessor or DSP, neither that the same self-test programs will obtain the same fault coverage in the real, commercial implementations of the same instructions set architectures.

Table 6-35 demonstrates the effectiveness of software-based self-testing on each of the selected benchmark processors that can be used in practical low-cost embedded systems. Table 6-35 summarizes for each benchmark: the processor size in gate equivalents, the functional components percentage with respect to the total processor area, the size of the self-test program, the execution time of the self-test program (without responses compaction) and the total single stuck-at fault coverage for the entire processor.

Benchmark Processor	Gate Count	Functional Components Percentage	Self-test program size	Execution Time (cycle)	Fault coverage
Plasma I	17,458	83.48%	965 w	3,552	92.2%
Plasma II	26,080	88.69%	853 w	5,797	95.3%
Plasma III	30,896	89.39%	853 w	5,797	94.5%
Plasma IV	27,824	89.26%	853 w	5,797	95.3%
Meister	37,402	68.99%	1,728 w	10,061	92.6%
Jam	43,208	66.85%	897 w	4,787	94.0%
oc8051	10,305	63.64%	3,760 b	5,411	93.1%
RISC-MCU	4,693	43.60%	1,258 w	2,446	91.2%
oc54x	11,611	92.21%	1,558 w	7,742	96.0%

Table 6-35: Summary of application of software-based self-testing.

From the contents of Table 6-35 we can draw the following conclusions that outline the effectiveness of software-based self-testing on the selected processor benchmarks.

- The self-test programs sizes are very small at the range of just a few kilobytes.
- The execution time of the self-test programs is, in all cases, less than 10,000 clock cycles.
- High fault coverage is obtained in all cases and the highest figures are obtained for the benchmarks that have a larger percentage of functional components.

Chapter 7

Processor-Based Testing of SoC

In this Chapter, we discuss the concept of software-based self-testing of SoC designs as an extension of software-based self-testing of embedded processors.

The basic idea of software-based self-testing of SoC is that an embedded processor that has already been tested by the execution of embedded software routines is subsequently used to perform testing of the remaining embedded cores of the SoC. Because of the central role of the embedded processor in SoC self-testing, the technique is also called processor-based self-testing.

In this Chapter first, we present the concept, some details and the advantages of software-based self-testing of SoC, and then we describe recent software-based self-testing approaches from the literature.

7.1 The concept

Software-based self-testing for SoC is shown in Figure 7-1 (this is an augmented version of Figure 2-1 where we show the basic mechanism of software-based self-testing of SoC). Software-based self-testing of SoC uses an embedded processor core supplied with self-test routines which have been downloaded from an external tester. The actual execution of testing is not hooked to this external tester but it is performed by the processor itself at the actual speed of operation of the SoC. The embedded processor core applies

Embedded Processor-Based Self-Test
D.Gizopoulos, A.Paschalis, Y.Zorian

tests to other cores of the SoC to which it has access and captures the core responses, playing the role of an "internal" tester.

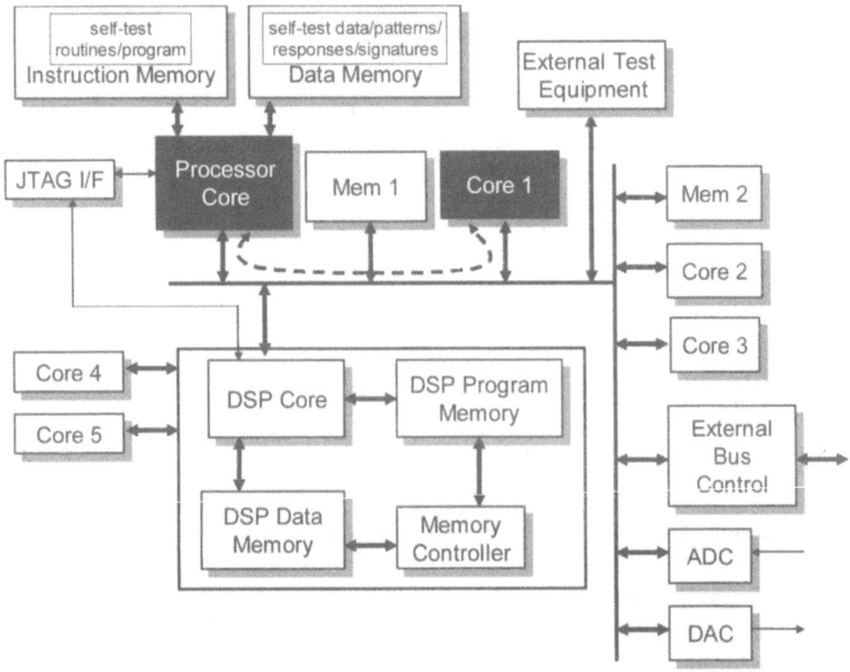

Figure 7-1: Software-based self-testing for SoC.

Self-testing a SoC using an embedded processor is not a straightforward task. Several aspects of this problem must be taken into consideration while different SoC architectures require different application of this generic strategy.

An embedded core in a SoC may be delivered from its IP provider supported with testing infrastructure (scan chains, DfT structures, P1500 wrapper) as well as a set of test patterns to be applied to it (when the core comes as a hard core). In this case, the embedded processor core must be supplied by the external test equipment with all necessary information (test patterns, protocol for their application) so that it can effectively apply the core test patterns and capture its test responses.

There may also be cases where an embedded core in a SoC is not supported by an existing test strategy. If the core is delivered in a gate level netlist form or a synthesizable HDL form (firm and soft core, respectively), then the core can be targeted for test generation either by a commercial ATPG tool (combinational or sequential) or by a pseudorandom test strategy. It may also be modified to include DfT structures (scan, test points, etc). The final testing approach of the core can be assigned for application to the processor core which will be supplied by the core test patterns and will effectively apply them to the core at the actual operating speed of the SoC (at-speed testing).

In other cases, an existing core may come with an *autonomous* testing mechanism, such as hardware-based self-testing. This is a very common situation in embedded memory cores which usually contain a memory BIST mechanism that applies a comprehensive memory test set to the core. Memory BIST mechanisms do not occupy excessive silicon area and most memory core providers deliver their cores with such embedded BIST facilities. In this case, the only role that the embedded processor is assigned in software-based self-testing of SoC, is to start and stop (if necessary) the memory BIST execution at the embedded memory core and capture the final signature of BIST for further evaluation. Availability of an autonomous mechanism in an embedded processor is not, of course, restricted only in the case of memories.

A serious factor that determines the efficiency of software-based self-testing in SoC designs is the mechanism that is used for the communication between the embedded processor and the external test equipment. Since this communication is restricted to the frequency of the external test equipment, it has an impact to the overall SoC test application time. If the self-test codes and self-test data/patterns/responses that must be transferred to the processor memory are downloaded with a low-speed serial interface, the total download time will me long and will seriously affect the overall test application time. Improvements can be seen if a parallel downloading mechanism is used, or even better if the external test equipment can communicate with the processor's memory using a Direct Memory Access (DMA) mechanism which does not interfere with the processor.

Apart from performing self-testing on the embedded cores of the SoC, testing can be performed by software-based self-testing to the SoC buses and interconnects. The overall performance of the SoC strongly depends on the performance of the interconnections and hence detection of defects/faults in these due to crosstalk must be achieved during manufacturing testing. The use of software-based self-testing for crosstalk testing has been studied in the literature as we will see in the next section.

Apart from the "as-is" use of an embedded processor for the self-testing of the cores and interconnects of a SoC, two other approaches may be used (have also been studied in the literature): (a) an existing embedded processor may be modified to include additional instructions (instruction-level DfT) to assist the testing task of the processor itself and the SoC overall; (b) a new processor, dedicated to SoC testing may be synthesized and used in the SoC just for the application of the SoC test strategy. In both scenaria (a) and (b) the objective is the reduction of the overall test application time of the SoC.

Both solutions may be very efficient if they do not add excessive area and performance overheads in the SoC architecture.

7.1.1 Methodology advantages and objectives

Software-based self-testing for SoC is an extension of software-based self-testing for a processor alone. In some sense, self-testing the SoC can be considered an "easier" problem than self-testing the processor itself because: (a) a processor is the most difficult design/core that is embedded in an SoC; (b) using a processor to test itself has the fundamental difficulty that the test pattern generator, the circuit under test and the test responses analyzer are *all* the same entity: the processor; in testing other SoC cores using the processor, there is a clear separation between the core that is being tested and the core that applied the test patterns and captures the test responses.

As the two problems (software-based self-testing for processor and for SoC) are closely related, they share the same list of advantage and have a common set of objectives. We briefly discuss about both in the following paragraphs.

Software-based self-testing for SoC is a *low-cost test strategy* which, on the other side, delivers *high test quality*. The advantages of software-based self-testing for SoC architectures are the following:

- *At-speed testing*. The SoC is tested at the actual frequency of operation and not in the – possibly lower – frequency of an external tester. Therefore, performance-related failure mechanisms (such as delay faults) can be also detected. These failure mechanisms escape detection when the chip is tested at lower frequencies.
- *Small hardware and performance overheads*. No additional circuit modifications are necessary other than those required by the particular testing strategy of each core. Even if instruction-level DfT is applied, the overall area and performance overhead to the entire SoC is very small, if it exists at all.
- *Yield improvement*. As already analyzed in the software-based self-testing analysis for processors, this methodology eliminates the yield loss which is due to external tester's inaccuracies. Being a test methodology that is applied by the chip itself, it reduces such inaccuracies and chip rejections related to them.
- *Flexibility, re-usability*. Software-based self-testing of SoC is a flexible self-testing methodology which gets the advantages of the flexibility of its main performer: the processor. For a SoC architecture, the software-based self-testing strategy can be revised at any stage to include new test sets for cores (for other

fault models[42] for example) without the need to change anything in the SoC design or test infrastructure. Moreover, an embedded processor can be re-used during the SoC life cycle to perform on-line/periodic testing and detect faults that appear after some usage of the system.

- *Low-cost testing.* Most of all, software-based self-testing for SoC is a low-cost test approach that is loosely hooked with expensive external test equipment. Extensive analysis of this aspect of the approach was given in previous Chapters.

SoC testing based on self-test routines execution on an embedded processor inherently possesses the previous advantages. In order to obtain a maximum of efficiency from the methodology, the following objectives must be satisfied by a particular software-based self-testing methodology for SoC.

- *Minimum interaction with external testers.* This objective, if met, leads to significant reduction of the SoC application time because the exchange of data between the SoC and tester is done at the frequency of the tester which is usually smaller than the SoC operating frequency.
- *Reduced self-test execution time.* This is a second objective that is related to SoC test application time. If a self-test routine for a core of the SoC is executed in less time, then the overall test application time for the core (and thus the SoC) is also reduced.
- *Small engineering effort.* This objective (obviously the ultimate objective of every testing approach) can be satisfied if software-based self-testing proceeds in two ways: (a) automation of self-test routines development; (b) targeting the processor cores that are most "important" for the total fault coverage of the SoC. The "importance" of a core in a SoC can be determined by its size compared with the size of other cores and other factors as well.

Software-based self-testing of processor-based SoC architectures is a recent direction in electronic testing which will gain importance as time passes. This is a prediction which is based on the increasing difficulty that SoC test engineers face today in multi-core, complex SoC architectures. The amount of test data that must be transferred from a tester to the SoC and vice versa is getting enormous and the direct impact that the use of expensive, high-memory, high-performance, high pin-count, testers has on the SoC

[42] The fault models used for the different cores of the SoC may significantly defer. While digital core testing is based on the single stuck-at fault model or a more comprehensive structural fault model such as the path delay fault model, memory cores may be tested for a variery of memory-related fault models.

manufacturing cost is recognized unanimously [76]. Solutions that will reduce the test costs associated with modern SoC designs and will also have a positive impact on yield improvement are requested. Software-based or processor-based self-testing is such a solution.

We present in the next section a brief discussion on recent works presented in the literature on SoC testing based on an existing embedded processor. Several different aspects of the problem have been tackled in these papers. Several others remain to be addressed by the methodologies of the future years.

7.2 Literature review

We don't use any particular order of the papers mentioned in this section, other than their chronological order, as we have already done in the processor testing literature review Chapter 4. The literature review of this section gives an idea of the current state of research in this field.

S.Hellebrand, H.-J.Wunderlich and A.Hertwig in 1996 [62] discussed a mixed-mode (combined pseudorandom and deterministic) self-testing scheme using embedded processors and software routines. A scan-based self-test scheme is the basis of this work. The approach tries to reduce the length of pseudorandom sequences combining them with appropriate deterministic test patterns in such a way that the memory requirements of the deterministic part are not excessive. Experimental results on several ISCAS circuits with software routines using Intel's 80960CA assembly language are reported.

J.Dreibelbis, J.Barth, H.Kalter and R.Kho proposed in 1998 [38] a built-in self-test architecture for embedded DRAMs based on a custom "processor-like" test engine. The test processor of this approach has two separate instruction memories and is designed so that the DRAM test dedicated pins are as few as possible. A reasonable area overhead for the test processor is reported.

R.Rajsuman in 1999 [134] presented an approach for SoC testing using an embedded microprocessor. At the first phase of the approach, the microprocessor is tested for the correct operation of all its instructions in a pseudorandom, functional BIST manner. Pseudorandom operands are generated by an LFSR and responses are compacted by a MISR. At the second phase, embedded memory cores are tested by software application of classic memory testing algorithms. At the third phase of the approach, other types of SoC cores are targeted. An example is given for the testing of an embedded Digital to Analog Converter (DAC). The approach is finally combined with an Iddq testing mechanism which is not related with the embedded microprocessor.

C.A.Papachristou, F.Martin and M.Nourani presented in 1999 [123] their approach for microprocessor-based testing of core-based SoC designs. The flexibility of processor-based SoC testing is outlined in this paper. A DMA mechanism is assumed for the transfer of test programs to the processor memory for subsequent execution by the processor. The access mechanism used to apply core tests is described. The approach has a pseudorandom philosophy with test patterns generated by LFSR and responses compacted in MISR. A detailed analysis of the entire SoC process (downloading, core access, test application, etc) is given and combined with a set of experimental results on a simple test SoC. A 96.75% fault coverage is reported for the simple SoC (5,792 faults in total).

A.Jas and N.A.Touba in 1999 [80] presented the use of an embedded processor for efficient compression/de-compression of test data in a SoC design. The SoC architecture is scan-based and the objective is the reduction of test data stored in tester memory as well as the over all test application time. Experimental results on ISCAS circuits demonstrate the improvements obtained from this approach.

X.Bai, S.Dey and J.Rajski proposed in 2000 [8] a self-test methodology to enable on-chip at-speed testing of crosstalk defects in SoC interconnects. The self-test methodology was based on the maximal aggressor fault model, that enables testing of the interconnect with a linear number of test patterns. Efficient embedded test generators that generate test vectors for crosstalk faults and embedded error detectors that analyze the transmission of the test sequences received from the interconnects and detect any transmission errors, were presented. Also test controllers to initiate and manage test transactions by activating appropriate test generators and error detectors, and having error diagnosis capability, were presented. The proposed self-test methodology was applied to test for crosstalk defects the buses of a DSP chip.

M.H.Tehranipour, Z.Navabi, and S.M.Fakhraie proposed in 2001 [153] a software-based self-test methodology that utilizes conventional microprocessors to test their on-chip SRAM. A mixture of existing memory testing techniques was adopted that covers all important memory faults. The derived test program implemented the "length 9N" memory test algorithm. This method can be implemented for embedded memory testing of all microprocessors, microcontrollers and DSPs. The proposed methodology was applied to the case of the 32K SRAM of the Texas Instruments TMS320C548 DSP.

J.-R.Huang, M.K.Iyer and K.-T.Cheng applied in 2001 [68] the concept of software-based self-testing on a bus-based SoC design. Cores are supplied with configurable test wrappers and self-test routines are executed by an embedded processor to test the cores. Fault coverage results are provided in the paper for the testing of ISCAS circuits used as example cores of the SoC.

C.-H.Tsai and C.-W.Wu proposed in 2001 [164] a processor-programmable built-in self-test scheme suitable for embedded memory testing in a SoC. The proposed self-test circuit can be programmed via an on chip microprocessor. Upon receiving the commands from the microprocessor, the test circuit generates pre-defined test patterns and compares the memory outputs with the expected outputs. Most popular memory test algorithms can be realized by properly programming the self-test circuit using the processor instructions. Compared with processor-based memory self-testing schemes that use a test program to generate test patterns and compare the memory outputs, the test time of the proposed memory BIST scheme was greatly reduced.

C.Galke, M.Pflanz and H.T.Vierhaus proposed in 2002 [49] the concept of designing a dedicated processor for the self-test of SoCs based on embedded processors. A minimum sized test processor was designed in order to perform on-chip test functions. Its architecture contained special adopted registers to realize LFSR or MISR functions for test pattern de-compaction and pattern filtering. The proposed test processor architecture is scalable and based on a standard RISC architecture in order to facilitate the use of standard compilers on it.

A.Krstic, W.-C.Lai, L.Chen, K.-T.Cheng and S.Dey presented in 2002 [101], [102] a review of the group's work in the area of software-based self-testing for processors and SoC designs. The software-based self-testing concept is presented in detail in these works and its advantages are clearly outlined. Discussion covers self-testing of the embedded processor (for stuck-at and delay faults), self-diagnosis of the embedded processor, self-testing of buses and interconnects, self-testing of other SoC cores using the processor and also instruction-level DfT. The authors have worked in these subtopics of software-based self-testing and a reference in more detail has been given in Chapter 4 for those that are related to processor self-testing only.

S.Hwang and J.A.Abraham discussed in 2002 [71] an optimal BIST technique for SoC using an embedded microprocessor. The approach aims to the reduction of memory requirements and test application time in scan-based core testing using pseudorandom and deterministic test patterns. This is achieve using a new test data compression technique using the embedded processor of the SoC. Experimental results are given using Intel's x86 instruction set (programs are developed in C and a compiler is used for their translation in assembly/machine language) on several ISCAS circuits. Comparisons are also reported with the previous works of [62], [80], and the superiority of the approach of [71] is shown in terms of the total test data size of the three approaches on the ISCAS circuits studied.

L.Chen, X.Bai, S.Dey proposed in 2002 [26] a new software-based self-test methodology for SoC based on embedded processors that utilizes an on-chip embedded processor to test at-speed the system-level interconnects for crosstalk. Testing of long interconnects for crosstalk in SoCs is important because crosstalk effects degrade the integrity of signals traveling on long interconnects. They demonstrated the feasibility of the proposed method by applying it to test the interconnects of a processor-memory system. The defect coverage was evaluated using a system-level crosstalk defect simulation environment.

M.H.Tehranipour, M.Nourani, S.M.Fakhraie and A.Afzali-Kusha in 2003 [154] outlined the use of embedded processor as a test controller in a SoC to test itself and the other cores. It is assumed that a DMA mechanism is available for efficient downloading of test programs in the processor memory. The flexibility of the approach of processor-based SoC testing is discussed. The approach is based on appropriate use of processor instructions that have access to the SoC cores. The approach is evaluated on a SoC design based on a DSP core called UTS-DSP which is compatible with TI's TMS320C54x DSP family. The SoC also contains an SRAM, a ROM, a Serial Port Interface and a Host Port Interface. A test program has been developed for the entire SoC consisting of a total of 689 bytes and a test execution time of 84.25 msec. The test program reached a 95.6% fault coverage for the DSP core, 100% for the two memories and 86.1% and 81.3% for the two interface cores, respectively. Although, in the work of [154], no sufficient details are given for the approach, the interest of this paper is that it gives an idea of the overall SoC testability that can be obtained by a very small embedded software program.

7.3 Research focus in processor-based SoC testing

Research in the area of processor-based or software-based self-testing of SoC architectures is expected to attract the interest of test researchers and engineers in the near future. Aspects of the approach that need special consideration are:

- Self-test *optimization* including: test application time reduction, test data volume reduction, compression/de-compression techniques of test data and parallelization of cores testing.
- *Diagnostic* capabilities of software-based self-testing for SoC. Performing diagnosis and identifying the malfunctioning parts of the SoC (cores or interconnects) will provide valuable

information for SoC manufacturing processes that will lead to yield improvements.

- *Automation* of self-test programs generation for different types of embedded SoC cores and different instruction sets of the embedded processors.
- *On-line testing* for SoC using an embedded processor as well as *fault tolerance* techniques utilizing embedded processors and software routines.

The advances on these topics will eventually prove the effectiveness of software-based self-testing as a generic, low-cost, high-test quality self-testing methodology for complex SoC architectures built around embedded processors.

Chapter 8

Conclusions

The emerging approach of software-based self-testing of embedded processors as well as processor-based SoC architectures has been discussed in this book. Software-based self-testing performs self-testing of processors and SoC based on the execution of embedded software routines, instead of assigning this task to dedicated hardware resources.

The definition of software-based self-testing sets as its essential objective the reduction of test costs for the processor and the SoC. Therefore, software-based self-testing is a *low-cost* test methodology. It does not add area or performance overheads to the design and it reduces the interaction with external test equipment which increases the test costs. Loose relation with external testers has another cost-related positive impact: yield loss due to the inherent tester's measurement inaccuracies is eliminated since testing is executed by the chip itself.

Moreover, software-based self-testing is a *high test quality* methodology because it allows the detection of failure mechanisms that can only be detected if the chip is tested at the actual operating frequency (at-speed testing).

Software-based self-testing is a *flexible* and *re-usable* self-testing technique that can be tuned to the specifics of a design, can be augmented for more comprehensive fault models and can also be re-used during the system's life cycle for on-line/periodic testing.

We have presented these aspects of the methodology in this book and we have described the requirements that must be satisfied as well as details on how the methodology objectives can be met.

By setting the requirements for efficient software-based self-testing, we want to set a common framework for comparisons in the topic. We discussed a flow for low-cost, software-based self-testing of embedded processors by tackling the most important components of the processor and by developing efficient self-test routines for each of them. Alternative styles in self-test routines development were discussed along with potential optimization techniques.

The software-based self-testing framework is also supported in a more quantitative manner, with a set of experimental results on publicly available embedded processor models of different complexities and architectures.

The future aims of research in the field of software-based self-testing for processors and SoC that will prove the long term usefulness of the approach and its applicability to a wide range of processor architectures are the following:

- Scalability: applicability of the approach to large, industrial embedded processors including several performance enhancing techniques.
- Automation: development of self-test routines for the components, self-test programs for the processor and the SoC in an automated flow that reduces test engineering effort and cost.

Finally, the application of software-based self-testing for on-line/periodic testing of processors and SoC, as well as for low-cost fault tolerance using embedded software routines will attract the interest of technologists.

References

[1] M.Abramovici, M.Breuer, A.D.Friedman, *Digital Systems Testing and Testable Design*, Piscataway, New Jersey: IEEE Press, 1994. Revised Printing.

[2] S.M.I.Adham, S.Gupta, "DP-BIST: a built-in self-test for DSP data paths-a low overhead and high fault coverage technique", Proceedings of the Fifth Asian Test Symposium (ATS) 1996, pp.205-512.

[3] V.D.Agrawal, C.R.Kime, K.K.Saluja, "A Tutorial on Built-In Self-Test, Part 1: Principles", IEEE Design & Test of Computers, vol. 10, no. 1, pp. 73-82, March 1993.

[4] V.D.Agrawal, C.R.Kime, K.K.Saluja, "A Tutorial on Built-In Self-Test, Part 2: Applications", IEEE Design & Test of Computers, vol. 10, no. 2, pp. 69-77, June 1993.

[5] D.Amason, A.L.Crouch, R.Eisele, G.Giles, M.Mateja, "A case study of the test development for the 2nd generation ColdFire^{R} microprocessors", Proceedings of the IEEE International Test Conference (ITC) 1997, pp. 424-432.

[6] M.Annaratone, M.G.Sami, "An Approach to Functional Testing of Microprocessors", Proceedings of the Fault Tolerant Computing Symposium (FTCS) 1982, pp. 158-164.

[7] P.J.Ashenden, J.P.Mermet, R.Seepold, *System-on-Chip Methodologies & Design Languages*, June 2001, Kluwer Academic Publishers.

[8] X.Bai, S.Dey, J.Rajski, "Self-Test Methodology for At-Speed Test of Crosstalk in Chip Interconnects", Proceedings of the ACM/IEEE Design Automation Conference (DAC) 2000, pp. 619-624.

[9] P.H.Bardell, W.H.McAnney, J.Savir, *Built-In Test for VLSI: Pseudorandom Techniques*, John Wiley and Sons, New York, 1987.

[10] K.Batcher, C.Papachristou, "Instruction randomization self test for processor cores", Proceedings of the IEEE VLSI Test Symposium (VTS) 1999, pp. 34-40.

[11] P.H.Bardell, M.J.Lapointe, "Production experience with Built-In Self-Test in the IBM ES/9000 System", Proceedings of the IEEE International Test Conference (ITC) 1991, pp. 28-37.

[12] P.G.Belomorski, "Pseudorandom self-testing of microprocessors", Microprocessing & Microprogramming, vol.19, no.1, pp. 37-47, January 1987.

[13] D.K.Bhavsar, D.R.Akeson, M.K.Gowan, D.B.Jackson, "Testability access of the high speed test features in the Alpha 21264 microprocessor", Proceedings of the IEEE International Test Conference (ITC) 1998, pp. 487-495.

[14] D.K.Bhavsar, J.H.Edmondson, "Testability strategy of the Alpha AXP 21164 microprocessor", Proceedings of the IEEE International Test Conference (ITC) 1994, pp. 50-59.

[15] D.K.Bhavsar, J.H.Edmondson, "Alpha 21164 testability strategy", IEEE Design & Test of Computers, vol.14, no.1, pp. 25-33, January-March 1997.

[16] D.K.Bhavsav, et al, "Testability Access of the High Speed Test Features in the Alpha 21264 Microprocessor", Proceedings of the IEEE International Test Conference (ITC) 1998, pp. 487-495.

[17] M.Bilbault, "Automatic testing of microprocessors", Electronique & Microelectronique Industrielles, no. 208, 1975, pp.31-33.

[18] U.Bieker, P.Marwedel, "Retargetable Self-Test Program Generation Using Constraint Logic Programming", Proceedings of the ACM/IEEE Design Automation Conference (DAC) 1995, pp. 605-611.

[19] P.E.Bishop, G.L.Giles, S.N.Iyengar, C.T.Glover, W.-O.Law, "Testability considerations in the design of the MC68340 Integrated Processor Unit", Proceedings of the IEEE International Test Conference (ITC) 1990, pp. 337-346.

[20] R.D.Blanton, J.P.Hayes, "Design of a Fast, Easily Testable ALU", Proceedings of the IEEE VLSI Test Symposium (VTS) 1996, pp. 9-16.

[21] D.Brahme, J.A.Abraham, "Functional Testing of Microprocessors", IEEE Transactions on Computers, vol. C-33, pp. 475-485, June 1984.

[22] R.L.Britton, *MIPS Assembly Language Programming*, Pearson Prentice Hall, Upper Saddle River, NJ, 2004.

[23] M.L.Bushnell, V.D.Agrawal, *Essentials of Electronic Testing for Digital, Memory and Mixed-Signal VLSI Circuits*, Kluwer Academic Publishers, 2000.

[24] A.Carbine, D.Feltham, "Pentium-Pro Microprocessor Design for Test and Debug", Proceedings of the IEEE International Test Conference (ITC) 1997.

[25] K.Chakrabarty, V.Iyengar, A.Chandra, *Test Resource Partitioning for System-on-a-Chip*, Kluwer Academic Publishers, 2002.

[26] L.Chen, X.Bai, S.Dey, "Testing for interconnect crosstalk defects using on-chip embedded processor cores", Journal of Electronic Testing: Theory and Applications, vol.18, no.4-5, pp. 529-538, August-October 2002.

[27] L.Chen, S.Dey, "DEFUSE: a Deterministic Functional Self-Test Methodology for Processors", Proceedings of the IEEE VLSI Test Symposium (VTS) 2000, pp. 255-262.

[28] L.Chen, S.Dey, "Software-Based Self-Testing Methodology for Processor Cores", IEEE Transactions on Computer Aided Design of Integrated Circuits and Systems, vol. 20, no. 3, pp. 369-380, March 2001.

[29] L.Chen, S.Dey, "Software-Based Diagnosis for Processors", Proceedings of the ACM/IEEE Design Automation Conference (DAC) 2002, pp. 259-262, June 2002.

[30] T.Cheng, E.Hoang, D.Rivera, A.Haedge, J.Fontenot, A.G.Carson, "Test Grading the 68332", Proceedings of the IEEE International Test Conference (ITC) 1991, pp. 150-159.

[31] L.Chen, S.Ravi, A.Raghunathan, S.Dey, "A Scalable Software-Based Self-Test Methodology for Programmable Processors", Proceedings of the ACM/IEEE Design Automation Conference (DAC) 2003, pp. 548-553, June 2003.

[32] F.Corno, M.Sonza Reorda, G.Squillero, M.Violante, "On the Test of Microprocessor IP Cores", Proceedings of the IEEE Design Automation & Test in Europe Conference (DATE) 2001, March 2001, pp.209-213.

[33] B.Courtois, "A Methodology for On-Line Testing of Microprocessors", Proceedings of the Fault-Tolerant Computing Symposium (FTCS) 1981, pp. 272-274.

[34] G.Crichton, "Testing microprocessors", IEEE Journal of Solid-State Circuits, June 1979, pp. 609-613.

[35] A.L.Crouch, M.Pressly, J.Circello, "Testability features of the MC68060 microprocessor", Proceedings of the IEEE International Test Conference (ITC) 1994, pp. 60-69.

[36] W.J.Culler, *Implementing Safety Critical Systems: The VIPER microprocessor*, Kluwer Academic Publishers, 1987.

[37] L.L.Day, P.A.Ganfield, D.M.Rickert, F.J.Ziegler, "Test Methodology for a Microprocessor with Partial Scan", Proceedings of the IEEE International Test Conference (ITC) 1998, pp. 708-716.

[38] J.Dreibelbis, J.Barth, H.Kalter, R.Kho, "Processor-Based Built-In Self-Test for Embedded DRAM", IEEE Journal of Solid-State Circuits, vol. 33, no. 11, pp. 1731-1740, November 1998.

[39] E.B.Eichelberger, E.Lindbloom, J.A.Waicukauski, T.W.Williams, *Structured Logic Testing*, Englewood Cliffs, New Jersey: Prentice-Hall, 1991.

[40] S.Erlanger, D.K.Bhavsar, R.Davies, "Testability Features of the Alpha 21364 Microprocessor", Proceedings of the IEEE International Test Conference (ITC) 2003, pp. 764-772.

[41] X.Fedi, R.David "Some experimental results from random testing of microprocessors", IEEE Transactions on Instrumentation & Measurement, vol. IM-35, no.1, pp. 78-86, March 1986.

[42] X.Fedi, R.David "Experimental results from Random testing of Microprocessors", Proceedings of the Fault-Tolerant Computing Symposium (FTCS) 1984, pp. 225-230.

[43] R.S.Fetherston, I.P.Shaik, S.C.Ma, "Testability features of AMD-K6™ microprocessor", Proceedings of the IEEE International Test Conference (ITC) 1997, pp. 406-413.

[44] T.G.Foote, D.E.Hoffman, W.V.Huott, T.J.Koprowski, B.J.Robbins, M.P.Kusko, "Testing the 400 MHz IBM generation-4 CMOS chip", Proceedings of the IEEE International Test Conference (ITC) 1997, pp. 106-114.

[45] J.F.Frenzel, P.N.Marinos, "Functional Testing of Microprocessors in a User Environment", Proceedings of the Fault Tolerant Computing Symposium (FTCS) 1984, pp. 219-224.

[46] R.A.Frohwerk, "Signature Analysis: A New Digital Field Service Method", Hewlett-Packard Journal, vol. 28, no. 9, pp. 2-8, May 1977.

[47] R.Fujii, J.A.Abraham, "Self-test for microprocessors", Proceedings of the International Test Conference (ITC) 1985, pp. 356-361.

[48] S.B.Furber, *ARM System-on-Chip Architecture* (2nd Edition), Addison-Wesley, August, 2000.

[49] C.Galke, M.Pflanz, H.T.Vierhaus, "A test processor concept for systems-on-a-chip", Proceedings of the IEEE International Conference on Computer Design (ICCD) 2002, pp.210-212.

[50] M.G.Gallup, W.Ledbetter, R.McGarity, S.McMahan, K.C.Scheuer, C.G.Shepard, L.Sood, "Testability features of the 68040", Proceedings of the IEEE International Test Conference (ITC) 1990, pp. 749-757.

[51] S.Ghosh, Hardware Description Languages: Concepts and Principles, New York: IEEE Press, 2000.

[52] D.Gizopoulos, A.Paschalis and Y.Zorian, "An Effective BIST Scheme for Datapaths", Proceedings of the IEEE International Test Conference, (ITC) 1996, pp. 76-85.

[53] D.Gizopoulos, A.Paschalis and Y.Zorian, "An Effective Built-In Self-Test Scheme for Booth Multipliers", IEEE Design & Test of Computers, vol. 15, no. 3, pp. 105-111, July-September 1998.

[54] D.Gizopoulos, A.Paschalis and Y.Zorian, "An Effective Built-In Self-Test Scheme for Array Multipliers", IEEE Transactions on Computers, vol. 48, no. 9, pp. 936-950, September 1999.

[55] D.Gizopoulos, A.Paschalis, Y.Zorian and M.Psarakis, "An Effective BIST Scheme for Arithmetic Logic Units", Proceedings of the IEEE International Test Conference, 1997, pp. 868-877.

[56] N.Gollakota, A.Zaidi, "Fault grading the Intel 80486", Proceedings of the IEEE International Test Conference (ITC) 1990, pp. 758-761.

[57] F.Golshan, "Test and On-line Debug Capabilities of IEEE Std 1149.1 in UltraSPARC™-III Microprocessor", Proceedings of the IEEE International Test Conference (ITC) 2000, pp. 141-150.

[58] A.J. van de Goor, Th.J.W. Verhallen, "Functional Testing of Current Microprocessors", Proceedings of the IEEE International Test Conference (ITC) 1992, pp. 684-695.

[59] M.Gumm, *VLSI Design Course: VHDL-Modelling and Synthesis of the DLXS RISC Processor*, University of Stuttgart, Germany, December 1995.

[60] R.Gupta, Y.Zorian, "Introducing Core-Based System Design", IEEE Design & Test of Computers, October-December 1997, pp. 15-25.

[61] K.Hatayama, K.Hikone, T.Miyazaki, H.Yamada, "A Practical Approach to Instruction-Based Test Generation for Functional Modules of VLSI Processors", Proceedings of the IEEE VLSI Test Symposium (VTS) 1997, pp. 17-22.

[62] S.Hellebrand, H.-J.Wunderlich, A.Hertwig, "Mixed-Mode BIST Using Embedded Processors", Proceedings of the IEEE International Test Conference (ITC) 1996, pp. 195-204.

[63] J.L.Hennessy, D.Patterson, *Computer Architecture – A Quantitative Approach*, Second Edition, San Francisco, CA: Morgan Kaufmann, 1996.

[64] B.Henshaw, "An MC68020 users test program", Proceedings of the International Test Conference (ITC) 1986, pp. 386-393.

[65] G.Hetherington, G.Sutton, K.M.Butler, T.J.Powell, "Test generation and design for test for a large multiprocessing DSP", Proceedings of the IEEE International Test Conference (ITC) 1995, pp. 149-156.

[66] K.Holdbrook, S.Joshi, S.Mitra, J.Petolino, R.Raman, M.Wong, "MicroSPARC: a case-study of scan based debug", Proceedings of the IEEE International Test Conference (ITC) 1994, pp. 70-75.

[67] H.Hong, R.Avra, "Structured design-for-debug-the SuperSPARC II methodology and implementation", Proceedings of the IEEE International Test Conference (ITC) 1995, pp. 175-183.

[68] J.-R. Huang, M. K. Iyer, K.-T. Cheng, "A Self-Test Methodology for IP Cores in Bus-Based Programmable SoCs", Proceedings of the IEEE VLSI Test Symposium (VTS) 2001, pp. 198-203.

[69] C.Hunter, E.K.Vida-Torku, J.LeBlanc, "Balancing structured and ad-hoc design for test: testing of the PowerPC 603 microprocessor", Proceedings of the IEEE International Test Conference (ITC) 1994, pp. 76-83.

[70] C.Hunter, J.Gaither, "Design and implementation of the "G2" PowerPC™ 603e-embedded microprocessor core", Proceedings of the IEEE International Test Conference (ITC) 1998, pp. 473-479.

[71] S.Hwang, J.A.Abraham, "Optimal BIST Using an Embedded Microprocessor", Proceedings of the IEEE International Test Conference (ITC) 2002, pp. 736-745.

[72] IEEE International Workshop on Test Resource Partitioning (TRP) 2000, 2001, 2002, 2003.

[73] IEEE P1500 SECT web site. http://grouper.ieee.org/groups/1500

[74] IEEE, *IEEE Standard Verilog Language Reference Manual*, Std 1364-1995, New York: IEEE, 1995 (also available in http://standards.ieee.org)

[75] IEEE, *IEEE Standard VHDL Language Reference Manual*, Std 1076-1993, New York: IEEE, 1993 (also available in http://standards.ieee.org)

[76] International Technology Roadmap for Semiconductors (ITRS), 2003 Edition, http://public.itrs.net.

[77] S.K.Jain, A.K.Susskind, "Test strategy for microprocessors", Proceedings of the ACM/IEEE 20th Design Automation Conference (DAC) 1983, pp.703-708.

[78] Jam CPU model, http://www.etek.chalmers.se/~e8mn/web/jam

[79] A.Jantsch, *Modeling Embedded Systems and SoC's: Concurrency and Time in Models of Computation*, Morgan Kaufmann, June 2003.

[80] A.Jas, N.A.Touba, "Using an embedded processor for efficient deterministic testing of systems-on-a-chip", Proceedings of the IEEE International Conference on Computer Design (ICCD) 1999, pp. 418-423.

[81] Y.Jen-Tien, M.Sullivan, C.Montemayor, P.Wilson, R.Evers, "Overview of PowerPC 620 multiprocessor verification strategy", Proceedings of the IEEE International Test Conference (ITC) 1995, pp. 167-174.

[82] J.Jishiura, T.Maruyama, H. Maruyama, S.Kamata, "Testing VLSI microprocessor with new functional capability", Proceedings of the IEEE International Test Conference (ITC) 1982, pp. 628-633.

[83] D.D.Josephson, D.J.Dixon, B.J.Arnold, "Test Features of the HP PA7100LC Processor", Proceedings of the IEEE International Test Conference (ITC) 1993, pp. 764-772.

[84] D.Josephson, S.Poehlman, V.Govan, C.Mumford, "Test Methodology for the McKinley Processor", Proceedings of the IEEE International Test Conference (ITC) 2001, pp. 578-585.

[85] G.Kane, J.Heinrich, *MIPS RISC Architecture*, Prentice Hall, 1992.

[86] M.G.Karpovsky, R.G. van Meter, "An approach to the testing of microprocessors", Proceedings of the ACM/IEEE 21st Design Automation Conference (DAC) 1984, pp.196-202.

[87] S.Karthik, M.Aitken, L.Martin, S.Pappula, B.Stettler, P.Vishakantaiah, M.d'Abreu, J.A.Abraham, "Distributed mixed level logic and fault simulation on the Pentium[R] Pro microprocessor", Proceedings of the IEEE International Test Conference (ITC) 1996, pp. 160-166.

[88] M.Keating, P.Bricaud, *Reuse Methodology Manual for System-on-a-Chip Designs*, Third Edition, June 2002, Kluwer Academic Publishers.

[89] A.Kinra, A.Mehta, N.Smith, J.Mitchell, F.Valente, "Diagnostic techniques for the UltraSPARC™ microprocessors", Proceedings of the IEEE International Test Conference (ITC) 1998, pp. 480-486.

[90] A.Kinra, "Towards Reducing 'Functional-Only' Fails for the UltraSPARC Microprocessors", Proceedings of the IEEE International Test Conference (ITC) 1999, pp. 147-154.

[91] H.-P.Klug, "Microprocessor Testing by Instruction Sequences Derived from Random Patterns", Proceedings of the IEEE International Test Conference (ITC) 1988, pp. 73-80.

[92] R.Koga, W.A.Kolasinski, M.T.Marra, W.A.Hanna, "Techniques of microprocessor testing and SEU-rate prediction", IEEE

Transactions on Nuclear Science, vol. NS-32, no.6, pp.4219-24, December 1985.

[93] N.Kranitis, D.Gizopoulos, A.Paschalis, M.Psarakis and Y.Zorian, "Power/Energy Efficient Built-In Self-Test Schemes for Processor Datapaths", IEEE Design & Test of Computers, vol. 17, no. 4, pp. 15-28, October-December 2000. Special Issue on Microprocessor Test and Verification.

[94] N.Kranitis, D.Gizopoulos, A.Paschalis, Y.Zorian, "Instruction-Based Self-Testing of Processor Cores", Proceedings of the IEEE VLSI Test Symposium (VTS) 2002, pp. 223-228.

[95] N.Kranitis, A.Paschalis, D.Gizopoulos, Y.Zorian, "Effective Software Self-Test Methodology for Processor Cores", Proceedings of the IEEE Design Automation and Test in Europe Conference (DATE) 2002, pp. 592-597.

[96] N.Kranitis, A.Paschalis, D.Gizopoulos, Y.Zorian, "Instruction-Based Self-Testing of Processor Cores", Journal of Electronic Testing: Theory and Applications, Kluwer Academic Publishers, Special Issue on VTS 2003, vol. 19, no 2, pp 103-112, April 2003.

[97] N.Kranitis, G.Xenoulis, D.Gizopoulos, A.Paschalis, Y.Zorian, "Low-Cost Software-Based Self-Testing of RISC Processor Cores", IEE Computers and Digital Techniques, Special Issue on DATE 2003 Conference, September 2003.

[98] N.Kranitis, G.Xenoulis, D.Gizopoulos, A.Paschalis, Y.Zorian, "Software-Based Self-Testing of Large Register Banks in RISC Processor Cores", Proceedings of the 4th IEEE Latin American Test Workshop 2003.

[99] N.Kranitis, Y.Xenoulis, D.Gizopoulos, A.Paschalis, Y.Zorian, "Low-Cost Software-Based Self-Testing of RISC Processor Cores", Proceedings of the IEEE Design Automation and Test in Europe Conference (DATE) 2003.

[100] N.Kranitis, G.Xenoulis, A.Paschalis, D.Gizopoulos, Y. Zorian, "Application and Analysis of RT-Level Software-Based Self-Testing for Embedded Processor Cores", Proceedings of the IEEE International Test Conference (ITC) 2003, pp. 431-440.

[101] A.Krstic, L.Chen, W.C.Lai, K.T.Cheng, S.Dey, "Embedded Software-Based Self-Test for Programmable Core-Based Designs", IEEE Design & Test of Computers, vol. 19, no. 4, pp. 18-26, July-August 2002.

[102] A.Krstic, W.C.Lai, L.Chen, K.T.Cheng, S.Dey, "Embedded Software-Based Self-Testing for SoC Design", Proceedings of the

ACM/IEEE Design Automation Conference (DAC) 2002, pp. 355-360.

[103] M.P.Kusko, B.J.Robbins, T.J.Snethen, P.Song, T.G.Foote, W.V.Huott, "Microprocessor test and test tool methodology for the 500 MHz IBM S/390 G5 chip", Proceedings of the IEEE International Test Conference (ITC) 1998, pp. 717-726.

[104] W.-C.Lai, K.-T.Cheng, "Instruction-Level DFT for Testing Processor and IP Cores in System-on-a-Chip", Proceedings of the ACM/IEEE Design Automation Conference (DAC) 2001, pp. 59-64, June 2001.

[105] W.-C.Lai, A.Krstic, K.-T.Cheng, "Test Program Synthesis for Path Delay Faults in Microprocessor Cores", Proceedings of the IEEE International Test Conference (ITC) 2000, pp. 1080-1089.

[106] P.K.Lala, "Microprocessor chip testing-a new method", Proceedings of the Microtest. Soc. Electronic & Radio Technicians, 1979, pp. 152-162.

[107] E.C.Lee, "A simple concept in microprocessor testing", Digest of Papers of the IEEE Semiconductor Test Symposium, 1976, pp. 13-15.

[108] J.Lee, J.H.Patel, "An Instruction Sequence Assembling Methodology for Testing Microprocessors", Proceedings of the IEEE International Test Conference (ITC) 1992, pp. 49-58.

[109] J.Lee, J.H.Patel, "Architectural Level Test Generation for Microprocessors", IEEE Transactions on Computer-Aided Design of Integrated Circuits and Systems, vol. 31, no. 10, pp. 1288-1300, October 1994.

[110] M.E.Levitt, S.Nori, S.Narayanan, G.P.Grewal, L.Youngs, A.Jones, G.Billus, S.Paramanandam, "Testability, debuggability, and manufacturability features of the UltraSPARC-I microprocessor", Proceedings of the IEEE International Test Conference (ITC) 1995, pp. 157-166.

[111] J.A.Lyon, M.Gladden, E.Hartung, E.Hoang, K.Raghunathan, "Testability Features of the 68HC16Z1", Proceedings of the IEEE International Test Conference (ITC) 1991, pp. 122-131.

[112] T.L.McLaurin, F.Frederick, "The Testability Features of the MCF5407 Containing the 4th Generation Coldfire Microprocessor Core", Proceedings of the IEEE International Test Conference (ITC) 2000, pp. 151-159.

[113] T.L.McLaurin, F.Frederick, R.Slobodnik, "The Testability Features of the ARM1026EJ Microprocessor Core", Proceedings of the IEEE International Test Conference (ITC) 2003, pp. 773-782.

[114] Meister Application Specific Instruction Processor, http://www.eda-meister.org

[115] B.T.Murray, J.P.Hayes, "Hierarchical Test Generation using Precomputed Tests for Modules", IEEE Transactions on Computer Aided Design of Integrated Circuits and Systems, vol. 19, no. 6, pp. 594-603, June 1990.

[116] Z.Navabi, *VHDL: Analysis and Modeling of Digital Systems*, New York: McGraw-Hill, 1993.

[117] W.Needham, N.Gollakota, "DFT strategy for Intel microprocessors", Proceedings of the IEEE International Test Conference (ITC) 1996, pp. 396-399.

[118] N.Nicolici, B.M.Al-Hashimi, *Power-Constrained Testing of VLSI Circuits*, Kluwer Academic Publishers, 2003.

[119] A.Noore, B.E.Weinrich, "Strategies for functional testing of microprocessors", Proceedings of the 22nd IEEE Southeastern Symposium on System Theory, 1990, pp.431-435.

[120] oc54x DSP model, http://www.opencores.org/projects/oc54x.

[121] oc8051 CPU model, http://www.opencores.org/projects/oc8051.

[122] Panel Session, "Microprocessor Testing: Which Technique is Best?", Proceedings of the ACM/IEEE Design Automation Conference (DAC) 1994, p. 294, June 1994.

[123] C.A.Papachristou, F.Martin, M.Nourani, "Microprocessor Based Testing for Core-Based System on Chip", Proceedings of the ACM/IEEE Design Automation Conference (DAC) 1999, pp. 586-591, June 1999.

[124] K.P.Parker, *The Boundary-Scan Handbook*, Third Edition, Kluwer Academic Publishers, 2003.

[125] P.Parvathala, K.Maneparambil, W.Lindsay, "FRITS – A Microprocessor Functional BIST Method", Proceedings of the IEEE International Test Conference (ITC) 2002, pp. 590-598.

[126] R.Patel, K.Yarlagadda, "Testability features of the SuperSPARC microprocessor", Proceedings of the IEEE International Test Conference (ITC) 1993, pp. 773-781.

[127] PicoJava Microprocessor Cores, Sun Microsystems [Online]. Available: http://www.sun.com/microelectronics/picoJava

[128] Plasma/MIPS CPU Model, http://www.opencores.org/projects/mips

[129] C.Pyron, J.Prado, J.Golab, "Next generation PowerPC™ microprocessor test strategy improvements", Proceedings of the IEEE International Test Conference (ITC) 1997, pp. 414-423.

[130] C.Pyron, M.Alexander, J.Golab, G.Joos, B.Long, R.Molyneaux, R.Raina, N.Tendolkar, "DFT Advances in the Motorola's MPC7400, a PowerPC™ G4 Microprocessor", Proceedings of the IEEE International Test Conference (ITC) 1999, pp. 137-146.

[131] K.Radecka, J.Rajski, J.Tyszer, "Arithmetic Built-In Self-Test for DSP Cores", IEEE Transactions on Computer Aided Design of Integrated Circuits and Systems, vol. 16, no. 11, pp. 1358-1369, November 1997.

[132] R.Raina, R.Bailey, D.Belete, V.Khosa, R.Molyneaux, J.Prado, A.Razdan, "DFT Advances in Motorola's Next-Generation 74xx PowerPC™ Microprocessor", Proceedings of the IEEE International Test Conference (ITC) 2000, pp. 132-140

[133] J.Rajski, J.Tyszer, *Arithmetic Built-In Self-Test for Embedded Systems*, Prentice-Hall, Upper Saddle River, New Jersey, 1998.

[134] R. Rajsuman, "Testing A System-on-Chip with Embedded Microprocessors", Proceedings of the IEEE International Test Conference (ITC) 1999, pp. 499-508.

[135] R.Regalado, "A 'people oriented' approach to microprocessor testing", Proceedings of the IEEE International Symposium on Circuits and Systems (ISCAS) 1975, pp. 366-368.

[136] RISC-MCU CPU model, http://www.opencores.org/projects/riscmcu/

[137] C.Robach, C.Bellon, G.Saucier, "Application oriented test for dedicated microprocessor systems", Microprocessors and their Applications, 1979, pp. 275-283.

[138] C.Robach, G.Saucier, "Application Oriented Microprocessor Test Method", Proceedings of the Fault Tolerant Computing Symposium (FTCS) 1980, pp. 121-125.

[139] C.Robach, G.Saucier, R.Velazco, "Flexible test method for microprocessors", Proceedings of the 6th EUROMICRO Symposium on Microprocessing and Microprogramming, 1980, pp.329-339.

[140] G.Roberts, J. Masciola, "Microprocessor boards and systems. A new approach to post in-circuit testing", Test, vol.6, no.2, pp. 32-34, March 1984.

[141] K.K.Saluja, L.Shen, S.Y.H.Su, "A Simplified Algorithm for Testing Microprocessors", Proceedings of the IEEE International Test Conference (ITC) 1983, pp. 668-675.

[142] K.K.Saluja, L.Shen, S.Y.H.Su, "A simplified algorithm for testing microprocessors", Computers & Mathematics with Applications, vol.13, no.5-6, 1987, pp.431-441.

[143] P.Seetharamaiah, V.R.Murthy, "Tabular mechanisation for flexible testing of microprocessors", Proceedings of the International Test Conference (ITC) 1986, pp. 394-407.

[144] J.Shen, J.Abraham, "Native Mode Functional Test Generation for Microprocessors with Applications to Self-Test and Design Validation", Proceedings of the International Test Conference (ITC) 1998, pp. 990-999.

[145] L.Shen, S.Y.H.Su, "A Functional Testing Method for Microprocessors", Proceedings of the Fault-Tolerant Computing Symposium (FTCS) 1984, pp. 212-218.

[146] L.Shen, S.Y.H.Su, "A Functional Testing Method for Microprocessors", IEEE Transactions on Computers, vol. 37, no. 10, pp. 1288-1293, October 1988.

[147] D.H.Smith, "Microprocessor testing – method or madness", Digest of Papers of the IEEE Semiconductor Test Symposium, 1976, pp. 27-29.

[148] J.Sosnowski, A.Kusmierczyk, "Pseudorandom versus Deterministic Testing of Intle 80x86 Processors", Proceedings of the IEEE Euromicro-22 Conference 1996, pp. 329-336.

[149] T.Sridhar, J.P.Hayes, "A Functional Approach to Testing Bit-Sliced Microprocessors", IEEE Transactions on Computers, vol. C-30, no. 8, pp. 563-571, August 1981.

[150] C.Stolicny, R.Davies, P.McKernan, T.Truong, "Manufacturing pattern development for the Alpha 21164 microprocessor", Proceedings of the IEEE International Test Conference (ITC) 1997, pp. 278-285.

[151] J.Sweeney, "Testability implemented in the VAX 6000 model 400", Proceedings of the IEEE International Test Conference (ITC) 1990, pp. 109-114.

[152] E.-S.A.Talkhan, A.M.H.Ahmed, A.E.Salama, "Microprocessors Functional Testing Techniques", IEEE Transactions on Computer-Aided Design, vol. 8, no. 3, pp. 316-318, March 1989.

[153] M.H.Tehranipour, Z.Navabi, S.M.Fakhraie, "An efficient BIST method for testing of embedded SRAMs", Proceedings of the IEEE International Symposium on Circuits and Systems (ISCAS) 2001, vol. 5, pp.73-76.

[154] M.H.Tehranipour, M.Nourani, S.M.Fakhraie, A.Afzali-Kusha, "Systematic Test Program Generation for SoC Testing Using Embedded Processor", Proceedings of the International Symposium on Circuits and Systems (ISCAS) 2003, pp. V541-V544.

[155] N.Tendolkar, B.Bailey, A.Metayer, B.Svrcek, E.Wolf, E.Fiene, M.Alexander, R.Woltenberg, R.Raina, "Test Methodology for Motorola's High-Performance e500 Core Based on PowerPC Instruction Set Architecture", Proceedings of the IEEE International Test Conference (ITC) 2002, pp. 574-583.

[156] S.M.Thatte, J.A.Abraham, "A Methodology for Functional Level testing of Microprocessors", Proceedings of the Fault-Tolerant Computing Symposium (FTCS) 1978, pp. 90-95.

[157] S.M.Thatte, J.A.Abraham, "Test generation for general microprocessors architectures", Proceedings of the Fault-Tolerant Computing Symposium (FTCS) 1979, pp. 203-210.

[158] S.M.Thatte, J.A.Abraham, "Test Generation of Microprocessors", IEEE Transactions on Computers, vol. C-29, pp. 429-441, June 1980.

[159] P.Thevenod-Fosse, R.David "Random testing of the Data Processing Section of a Microprocessor", Proceedings of the Fault-Tolerant Computing Symposium (FTCS) 1981, pp. 275-280.

[160] P.Thevenod-fosse, R.David, "Random Testing of Control Section of a Microprocessor", Proceedings of the Fault Tolerant Computing Symposium (FTCS) 1983, pp. 366-373.

[161] C.Timoc, F.Stoot, K.Wickman, L.Hess, "Adaptative self-test for a microprocessor", Proceedings of the IEEE International Test Conference (ITC) 1983, pp. 701-703.

[162] Q.Tong, N.K.Jha, "Design of C-testable DCVS binary array dividers", IEEE Journal of Solid-State Circuits, vol. 26, no. 2, pp. 134-141, February 1991.

[163] O.A.Torreiter, U.Baur, G.Goecke, K.Melocco, "Testing the enterprise IBM System/390™ multi processor", Proceedings of the IEEE International Test Conference (ITC) 1997, pp. 115-123.

[164] C-H.Tsai, C-W.Wu. "Processor-programmable memory BIST for bus-connected embedded memories", Proceedings of the Asia and South Pacific Design Automation Conference (ASP-DAC) 2001, pp. 325-330.

[165] R.S.Tupuri, J.A.Abraham, "A Novel Functional Test Generation Method for Processors using Commercial ATPG", Proceedings of the IEEE International Test Conference (ITC) 1997, pp. 743-752.

[166] R.S.Tupuri, A.Krishnamachary, J.A.Abraham, "Test Generation for Gigahertz Processor Using an Automatic Functional Constraint Extractor", Proceedings of the ACM/IEEE Design Automation Conference (DAC) 1999, pp. 647-652, June 1999.

[167] G.Vandling, "Modeling and Testing the Gekko Microprocessor, An IBM PowerPC Derivative for Nintendo", Proceedings of the IEEE International Test Conference (ITC) 2001, pp. 593-599.

[168] R.Velazco, H.Ziade, E.Kolokithas, "A microprocessor test approach allowing fault localisation", Proceedings of the International Test Conference (ITC) 1985, pp.737-743.

[169] B.Williams, "LSI automatic test equipment applied to dynamic microprocessor testing", New Electronics, vol. 7, no. 21, 1974, pp. 50-52.

[170] M.J.Y.Williams, J.B.Angell, "Enhancing Testability of Large-Scale Integrated Circuits via Test Points and Additional Logic", IEEE Transactions on Computers, vol. C-22, no. 1, pp. 46-60, January 1973.

[171] W.Wolf, *Modern VLSI Design: System-on-Chip Design*, 3rd Edition, Prentice Hall, January 2002.

[172] T.Wood, "The Test and Debug Features of the AMD-K7 Microprocessor", Proceedings of the IEEE International Test Conference (ITC) 1999, pp. 130-136.

[173] G.Xenoulis, D.Gizopoulos, N.Kranitis, A.Paschalis, "Low-Cost On-Line Software-Based Self-Testing of Embedded Processor Cores", Proceedings of the 9th IEEE International On-Line Testing Symposium (IOLTS) 2003, pp. 149-154.

[174] Xtensa™ Microprocessor Overview Handbook, Tensilica Inc., http://www.tensilica.com/xtensa_overview_handbook.pdf, August 2001

[175] W.Zhao, C.Papachristou, "Testing DSP Cores Based on Self-Test Programs", Proceedings of the IEEE Design Automation & Test in Europe Conference (DATE) 1998, pp. 166-172.

[176] Y.Zorian, "A distributed BIST control scheme for complex VLSI devices", Proceedings of the IEEE VLSI Test Symposium (VTS) 1993, pp. 4-9.

Index

About the Authors

Dimitris Gizopoulos is an Assistant Professor at the Department of Informatics, University of Piraeus, Greece. His research interests include processor testing, design-for-testability, self-testing, on-line testing and fault tolerance of digital circuits. Gizopoulos received the Computer Engineering degree from the University of Patras, Greece and a PhD from the University of Athens, Greece. He is author of more than sixty technical papers in transactions, journals, books and conferences and co-inventor of a US patent. He is member of the editorial board of IEEE Design & Test of Computers Magazine, and guest editor of special issues in IEEE publications. He is a member of the Steering, Organizing and Program Committees of several test technology technical events, member of the Executive Committee of the IEEE Computer Society Test Technology Technical Council (TTTC), a Senior Member of the IEEE and a Golden Core Member of the IEEE Computer Society.

Antonis Paschalis is an Associate Professor at the Department of Informatics and Telecommunications, University of Athens, Greece. Previously, he was Senior Researcher at the Institute of Informatics and Telecommunications of the National Research Centre "Demokritos" in Athens. He holds a B.Sc. degree in Physics, a M.Sc. degree in Electronics and Computers, and a Ph.D. degree in Computers, all from University of Athens. His current research interests are logic design and architecture, VLSI testing, processor testing and hardware fault-tolerance. He has published over 100 papers and holds a US patent. He is a member of the editorial board of JETTA and has served the test community as vice chair of the Communications Group of the IEEE Computer Society TTTC and participating in several organizing and program committees of international events in the area of design and test.

Yervant Zorian is the Vice President and Chief Scientist of Virage Logic Corp. Previously he was the Chief Technology Advisor of LogicVision Inc. and a Distinguished Member of Technical Staff at AT&T Bell Laboratories. Zorian received the MSc degree in Computer Engineering from the University of Southern California and a PhD in electrical engineering from McGill University. He also holds an executive MBA from Warthon School of Business, University of Pennsilvenia. He is the author of over 200 technical papers and three books, has received several best paper awards and holds twelve U.S. patents. Zorian serves as the IEEE Computer Society's Vice President for Technical Activities and the Editor-in-Chief Emeritus of the IEEE Design & Test of Computers. He participates in editorial advisory boards of IEEE Spectrum, and JETTA. He chaired the Test Technology Technical Council of IEEE Computer Society, and founded IEEE P1500 Standard Working Group. He is a Golden Core Member of IEEE Computer Society, Honorary Doctor of National Academy of Sciences of Armenia, and a Fellow of the IEEE.

MIX
Papier aus verantwortungsvollen Quellen
Paper from responsible sources
FSC® C105338

If you have any concerns about our products,
you can contact us on
ProductSafety@springernature.com

In case Publisher is established outside the EU,
the EU authorized representative is:
Springer Nature Customer Service Center GmbH
Europaplatz 3, 69115 Heidelberg, Germany

Printed by Libri Plureos GmbH
in Hamburg, Germany